Studies in Applied Philosophy, Epistemology and Rational Ethics

Volume 21

Series editor

Lorenzo Magnani, University of Pavia, Pavia, Italy
e-mail: lmagnani@unipv.it

Editorial Board

Atocha Aliseda
Universidad Nacional Autónoma de México (UNAM), Coyoacan, Mexico

Giuseppe Longo
Centre Cavaillès, CNRS—Ecole Normale Supérieure, Paris, France

Chris Sinha
Lund University, Lund, Sweden

Paul Thagard
Waterloo University, Waterloo, ON, Canada

John Woods
University of British Columbia, Vancouver, BC, Canada

About this Series

Studies in Applied Philosophy, Epistemology and Rational Ethics (SAPERE) publishes new developments and advances in all the fields of philosophy, epistemology, and ethics, bringing them together with a cluster of scientific disciplines and technological outcomes: from computer science to life sciences, from economics, law, and education to engineering, logic, and mathematics, from medicine to physics, human sciences, and politics. It aims at covering all the challenging philosophical and ethical themes of contemporary society, making them appropriately applicable to contemporary theoretical, methodological, and practical problems, impasses, controversies, and conflicts. The series includes monographs, lecture notes, selected contributions from specialized conferences and workshops as well as selected Ph.D. theses.

Advisory Board

A. Abe, Chiba, Japan
H. Andersen, Copenhagen, Denmark
O. Bueno, Coral Gables, USA
S. Chandrasekharan, Mumbai, India
M. Dascal, Tel Aviv, Israel
G.D. Crnkovic, Västerås, Sweden
M. Ghins, Lovain-la-Neuve, Belgium
M. Guarini, Windsor, Canada
R. Gudwin, Campinas, Brazil
A. Heeffer, Ghent, Belgium
M. Hildebrandt, Rotterdam,
    The Netherlands
K.E. Himma, Seattle, USA
M. Hoffmann, Atlanta, USA
P. Li, Guangzhou, P.R. China
G. Minnameier, Frankfurt, Germany
M. Morrison, Toronto, Canada
Y. Ohsawa, Tokyo, Japan
S. Paavola, Helsinki, Finland
W. Park, Daejeon, South Korea

A. Pereira, São Paulo, Brazil
L.M. Pereira, Caparica, Portugal
A.-V. Pietarinen, Helsinki, Finland
D. Portides, Nicosia, Cyprus
D. Provijn, Ghent, Belgium
J. Queiroz, Juiz de Fora, Brazil
A. Raftopoulos, Nicosia, Cyprus
C. Sakama, Wakayama, Japan
C. Schmidt, Le Mans, France
G. Schurz, Dusseldorf, Germany
N. Schwartz, Buenos Aires, Argentina
C. Shelley, Waterloo, Canada
F. Stjernfelt, Aarhus, Denmark
M. Suarez, Madrid, Spain
J. van den Hoven, Delft,
    The Netherlands
P.-P. Verbeek, Enschede,
    The Netherlands
R. Viale, Milan, Italy
M. Vorms, Paris, France

More information about this series at http://www.springer.com/series/10087

Susann Wagenknecht · Nancy J. Nersessian
Hanne Andersen
Editors

Empirical Philosophy of Science

Introducing Qualitative Methods into Philosophy of Science

Editors
Susann Wagenknecht
Centre for Science Studies
Aarhus University
Aarhus
Denmark

Hanne Andersen
Department for Science Education
University of Copenhagen
Copenhagen
Denmark

Nancy J. Nersessian
Department of Psychology
Harvard University
Cambridge, MA
USA

ISSN 2192-6255   ISSN 2192-6263 (electronic)
Studies in Applied Philosophy, Epistemology and Rational Ethics
ISBN 978-3-319-18599-6   ISBN 978-3-319-18600-9 (eBook)
DOI 10.1007/978-3-319-18600-9

Library of Congress Control Number: 2015939998

Springer Cham Heidelberg New York Dordrecht London
© Springer International Publishing Switzerland 2015
This work is subject to copyright. All rights are reserved by the Publisher, whether the whole or part of the material is concerned, specifically the rights of translation, reprinting, reuse of illustrations, recitation, broadcasting, reproduction on microfilms or in any other physical way, and transmission or information storage and retrieval, electronic adaptation, computer software, or by similar or dissimilar methodology now known or hereafter developed.
The use of general descriptive names, registered names, trademarks, service marks, etc. in this publication does not imply, even in the absence of a specific statement, that such names are exempt from the relevant protective laws and regulations and therefore free for general use.
The publisher, the authors and the editors are safe to assume that the advice and information in this book are believed to be true and accurate at the date of publication. Neither the publisher nor the authors or the editors give a warranty, express or implied, with respect to the material contained herein or for any errors or omissions that may have been made.

Printed on acid-free paper

Springer International Publishing AG Switzerland is part of Springer Science+Business Media (www.springer.com)

Contents

Empirical Philosophy of Science: Introducing Qualitative Methods into Philosophy of Science 1
Susann Wagenknecht, Nancy J. Nersessian and Hanne Andersen

Part I Foundations

Prolegomena to an Empirical Philosophy of Science 13
Lisa M. Osbeck and Nancy J. Nersessian

Feeling with the Organism: A Blueprint for an Empirical Philosophy of Science. 37
Erika Mansnerus and Susann Wagenknecht

Part II Case Studies

Modeling as a Case for the Empirical Philosophy of Science 65
Ekaterina Svetlova

Reductionism as an Identity Marker in Popular Science 83
Hauke Riesch

An Empirical Method for the Study of Exemplar Explanations 105
Mads Goddiksen

Longino's Theory of Objectivity and Commercialized Research. 127
Saana Jukola

Part III Empirical Philosophy of Science and HPS

History and Philosophy of Science as an Interdisciplinary Field of Problem Transfers................................... 147
Henrik Thorén

Context-Dependent Anomalies and Strategies for Resolving Disagreement.. 161
Douglas Allchin

Empirical Philosophy of Science: Introducing Qualitative Methods into Philosophy of Science

Susann Wagenknecht, Nancy J. Nersessian and Hanne Andersen

Abstract A growing number of philosophers of science make use of qualitative empirical data, a development that may reconfigure the relations between philosophy and sociology of science and that is reminiscent of efforts to integrate history and philosophy of science. Therefore, the first part of this introduction to the volume *Empirical Philosophy of Science* outlines the history of relations between philosophy and sociology of science on the one hand, and philosophy and history of science on the other. The second part of this introduction offers an overview of the papers in the volume, each of which is giving its own answer to questions such as: Why does the use of qualitative empirical methods benefit philosophical accounts of science? And how should these methods be used by the philosopher?

Keywords Empirical methods · Qualitative research · History and philosophy of science · Naturalized philosophy

---

S. Wagenknecht (✉)
Center for Science Studies, Aarhus University, Ny Munkegade 118,
8000 Aarhus C, Denmark
e-mail: Su.wagen@gmail.com

N.J. Nersessian
Department of Psychology, Harvard University, 1160 William James Hall,
33 Kirkland St., Cambridge 02138, MA, USA
e-mail: nancyn@cc.gatech.edu

H. Andersen
Department for Science Education, Øster Voldgade 3,
1350 Copenhagen K, Denmark
e-mail: hanne.andersen@ind.ku.dk

1 Introduction

When philosophers of science make use of qualitative methods, they draw upon a long and rich research tradition that is rooted in the social sciences and has gradually been adopted in other fields.[1] The use of qualitative methods in philosophy of science brings philosophers in close contact with philosophically inclined social scientists studying science, and at the same time it brings forth new perspectives on the classical problem of the integration of history and philosophy of science. This introduction will give an overview of the new relations to sociology of science and history of science brought about by the use of qualitative methods, and it will shortly present the papers in the volume—each of which gives its own answer to two questions: Why does the use of qualitative empirical methods benefit philosophical accounts of science? And how should these methods be used by the philosopher?

1.1 History and Philosophy of Science

The role that empirical insight can play in the philosophy of science is extensively debated in discussions on the relevance of historical to philosophical accounts of science (and vice versa). The relation between history and philosophy of science has been an issue of contention for half a century.[2] In the late 1950s and early 1960s, a new historiography of science developed that aimed at describing past science in its historical integrity rather than through the lens of the present, and by doing so it provided an image of science that seemed to differ from the image entailed by standard philosophical accounts at the time. Historically inclined philosophers of science therefore began suggesting that philosophy of science should be concerned with the historical structure of science rather than with an ahistorical, logical structure that they saw as having little relation to the actual scientific enterprise. In addition, they advocated that philosophers should conduct their own historical research directed towards specifically philosophical questions rather than rely on accounts developed by historians.

---

[1]While qualitative methods have gradually been adopted by many fields outside sociology, the methods themselves have also developed (for histories of how qualitative methods have developed and been received, see Denzin and Lincoln (2000), Vidich and Lyman (1994), Brinkmann et al. (2014)). The idea of qualitative methods as it is used today was established during the 1960s and 1970s when the first handbooks, textbooks and specialized journals focused on qualitative methods appeared (see e.g. Glaser and Strauss 1967; Filstead 1970; Bogdan and Taylor 1975). During the 1980s, they were increasingly adopted within psychology, educational research and areas such as nursing science, and a decade later in health care research more generally.

[2]We are here referring primarily to the Anglo-Saxon tradition of history and philosophy of science. In the continental tradition, the history of the relation has been different, see e.g. Gutting (1989, 2005). For overviews of the Anglo-Saxon history, see e.g. Mauskopf and Schmaltz (2012).

In attending to science through history, this emerging historical philosophy of science was faced with the question whether the accounts it provided were descriptive or prescriptive. In addressing this question Kuhn, for example, argued that his theses about the structure of scientific revolutions should be read in both ways at once: his account of the development of science was a prescriptive theory, and the reasons for taking it seriously were that scientists do in fact behave as the theory says they should. This might seem circular, but as with any other theory, the success of a theory of the development of science should be dependent on its ability to explain new data that had not been involved in its initial formulation.

Parallel to the growing institutionalization of history and philosophy of science (HPS) as a field, the discussion continued whether the relation between history and philosophy of science really was an intimate relationship, or if it was rather just a marriage of convenience (see e.g. Feigl 1970; McMullin 1970; Giere 1973; Burian 1977).[3] Whereas most philosophers of science agreed that philosophy of science had to be informed by a close attention to science, there was less agreement on whether the history of science was to play a privileged role. Arguments in favor of an integrated history and philosophy of science varied from the more pragmatic argument that early science is often more accessible than contemporary science to more principled arguments, asserting that topics of a particular type, such as how science develops over time, necessarily require a historical perspective.

Another topic of discussion was how historical cases could and should be selected and what and how philosophers could generalize from them. On the one hand, philosophers critical of the historical turn argued that if cases were selected to illustrate a philosophical position already developed, then it could be questioned as to how far these cases would work as support. Conversely, if starting from the historical cases, it was questionable how much could be generalized from just a few or sometimes even a single case (see e.g. Pitt 2001; Burian 2001). On the other hand, historically inclined philosophers of science argued that history of science neither generated facts from which philosophical generalization could be induced, nor did it generate evidence by which philosophical theories could be directly tested. Instead, historical cases and philosophical analyzes need to be integrated in a mutual, iterative process (Chang 2012).

1.2 Beyond History: A Broader Approach to Naturalized Philosophy

In the following decades, the historically inclined philosophy of science came to be seen as just one approach of a *naturalized* philosophy of science. Following on Quine's (1969) argument for "naturalizing" epistemology by using findings from

---

[3]See also Nickles (1995), Schickore (2011) and the collection of papers in Mauskopf and Schmaltz (2012) for later surveys of the debate.

scientific investigations from biology, psychology, and sociology to advance the project of justification of knowledge claims, a number of philosophers of science advocated naturalistic approaches as well (see, e.g. Giere 1988; Goldman 1986; Kitcher 1993; Laudan 1977; Nersessian 1984; Thagard 1988).

Quine had argued that justification and the status of knowledge claims depend on the characteristics of processes that generate and maintain belief. One approach was to inform analyses of science with findings and theories from the rapidly growing cognitive sciences on how humans in general perceive and generate knowledge about the world (Barker et al. 2001; Chen et al. 1998; Andersen et al. 1996; Darden 1991; Giere 1988; Gooding 1990; Nersessian 1984, 1987, 1992, 1995). Nersessian (1987) coined the term 'cognitive-historical' analysis for the philosophical method that took into account the cognitive processes of scientists in their construction of knowledge. In defense against the charge of 'circularity' in using the findings of science to study science, she argued: "The assumptions, methods, and results from both sides are subjected to critical evaluation, with corrective insights moving in both directions. The goal is to bring historical and cognitive interpretations into a state of reflective equilibrium, so as to make the circularity inherent in the approach virtuous rather than vicious." (Nersessian 1995, p. 196).

1.3 Sociology and Philosophy of Science

Whereas the relation between history and philosophy of science has been seen as a marriage, although it was up for dispute whether this marriage was established by love or convenience, the relation between sociology and philosophy of science has varied from periods of polite indifference to periods of mutual hostile competition.

At the time when HPS emerged and institutionalized, sociology of science was dominated by scholars such as Merton (1973), Hagstrom (1965), Zuckerman (1978), Chubin (1976), Cole and Cole (1973) who all took a macrosociological approach focused on describing the social structure and culture of science, including norms and deviant behavior, stratification and reward, and the growth and decline of scientific specialties. The work on scientific specialties, in particular, had obvious relations to Kuhn's work on paradigms and normal science, and Kuhn himself explicitly referred to the work of Hagstrom and others as the key to identifying scientific communities. This early macrosociology of science included both quantitative and qualitative studies, but with the emergence of the Science Citation Index in the 1960s quantitative studies of citations and co-citations gained prominence and eventually developed into its own specialty of scientometrics (see Wouters 1999, especially Chap. 4).

During the 1970s and 1980s sociologists of science increasingly turned away from the macrosociological focus on scientific communities and their stratification, and turned instead towards a microsociological focus on scientists' practices as they unfold locally in the laboratory. Key contributions to this new microsociology, such as Latour and Woolgar's (1979) *Laboratory Life* or Knorr-Cetina's (1981)

The Manufacture of Knowledge, showed how ethnographic and other qualitative methods could be used to study scientific practice. At the same time, in arguing for a social constructivism on which it should be explained in purely sociological terms why scientists believe what they do and how scientific ideas, methods and practices change over time, the more radical versions of this new microsociology of science crossed philosophers' favored boundary between internal and external aspects of science. The reactions were seen in the Science Wars of the 1990s, when both philosophers and scientists attacked what they saw as a dangerous attack on rationality. Hence, although the new microsociology of science devised an empirical method for how detailed case studies could be made of the practices of contemporary science in contrast to historical science, philosophers (at least in the Anglo-Saxon tradition), did not initially see it as opening up new venues for philosophical investigations that could supplement historical case studies.

1.4 The Latest Turns in Philosophy of Science

In recent decades, philosophers of science of various inclinations have become more interested in sociology of science and in including qualitative methods in their philosophical repertoire. Cognitively inclined philosophers of science have tried to overcome the boundary between the cognitive and the social through ethnographic investigations of distributed cognition in order, for example, to understand the reasoning practices in the evolving cognitive-cultural systems that comprise a modern research laboratory staffed with scientists with different competences and equipped with a multitude of artifacts (Nersessian 2006). Other philosophers of science have started pointing out that epistemologists' traditional focus on the warranted/justified beliefs of the individual cognitive agent leaves out important social aspects in scientific knowledge creation and have used qualitative research methodologies to investigate such topics as epistemic dependence and the dynamics of epistemic trust (Wagenknecht 2014, 2015). The use of qualitative methods to investigate philosophical topics has also spread beyond topics related to the social aspects of knowledge production, for example to investigate scientists' views on models and modelling (Bailer-Jones 2002; Chandrasekharan and Nersessain 2015; Mattila 2005; MacLeod and Nersessian 2013; Nersessian and Patton 2009), simplicity (Riesch 2010), or conceptions of risk (Mansnerus 2012).

Much of this work mirrors the empirical turn that has been seen in ethics from the 1990s onwards, especially in medical and bioethics where there is a growing literature reporting empirical investigations of people's actual moral intuitions, beliefs, reasoning and behavior by means of qualitative methods such as interviews or participant observation. Like in philosophy of science, a recurrent theme in this literature has been the relation between the descriptive and the normative, and the means by which philosophical inquiry and empirical research can be fruitfully integrated and contribute to the development of normative ethics (see e.g. Kon 2009; Leget et al. 2009; de Vries and van Leeuwen 2010).

2 Overview of This Volume

The papers in this volume explore benefits and challenges of an empirical philosophy of science and address questions such as: What do philosophers gain from empirical work? How can empirical research help to develop philosophical concepts? How do we integrate philosophical frameworks and empirical research? What constraints do we accept when choosing an empirical approach? What constraints does a pronounced theoretical focus impose on empirical work?

2.1 Part I: "Foundations"

The first part of this volume lays out the foundations of what an empirical philosophy of science could be. In two papers, philosophers of science reflect upon their use of qualitative empirical methods such as participant observation and open-ended interviewing. These reflections address basic issues of the relationship that empirical philosophy of science creates between first-hand, empirical insight and philosophical theorizing.

In addressing the delineation of the empirical from the non-empirical, *Osbeck and Nersessian* confront a fundamental issue in the use of empirical methods in philosophy of science. In the first part of their paper, the authors point out that non-empirical questions are key to any empirically-based analysis. Based on their experience with the formulation of empirically-based, philosophical accounts of science, the authors elaborate on two such non-empirical questions: What counts as an empirical approach to the study of science? And, given an empirical approach has been chosen, what is the appropriate unit of analysis for such a study? Osbeck and Nersessian show that each of the two questions opens a range of possibilities and requires a series of well-argued for choices. When non-empirical questions are settled, empirical questions can be tackled. Drawing on their own research, the authors show which philosophically relevant issues can be fruitfully approached as empirical questions, and they describe in instructive detail the qualitative empirical methods that they have previously employed and the insights that they were able to gain.

Mansnerus and Wagenknecht tackle the link between empirical and non-empirical (or, as they refer to it, between concrete and abstract) from a different angle. Reflecting upon the experiences that they as philosophers have had with the use of qualitative methods, they describe the relationship between philosophical conceptualization and empirical data as an iterative dialogue between abstract and concrete, theory and data. In their view, this dialogue benefits from a 'feeling with' the empirical phenomenon under study that the philosopher-investigator is able to develop in the course of her fieldwork. The authors describe how the dialogical interplay between conceptual discourse and concrete empirical insight manifests itself in their work, i.e., when analyzing the practices of infectious disease modelling or, respectively, studying a team of planetary scientists.

2.2 Part II: "Case Studies"

The second part of this volume compiles four papers that argue for the use of empirical knowledge, gained through qualitative empirical methods, in the philosophical analysis of scientific practice. These papers span problems as varied as the epistemic character of modeling practices in finance, the representation of reductionism in popular science, the study of explanations in science textbooks, and the investigation of commercialized biomedical research. Except for the last paper in this part, all authors provide hands-on insights gained from personal experience with the use of diverse qualitative methods ranging from ethnographic methods to text analysis.

Based on her experience with the study of modeling practice in finance as a participant observer, *Svetlova* assesses the challenges and gains that an empirically-based philosophy of science faces. To do so, she begins by revisiting the debate about issues associated with "experimental philosophy", a mode of philosophical inquiry that makes use of quantitative methods such as surveying. She then discusses whether and to what extent philosophical approaches, including her own, face similar issues when they employ qualitative methods such as participant observation, in-depth interviews and text-analysis. For the study of modeling practice, the author argues, empirical approaches have proven fruitful, and she provides three concrete ways in which empirical insights have furthered her philosophical theorizing on the nature of scientific models: by giving a new inspirational impulse to philosophical theorizing, by challenging an existing theoretical point of view, and by providing background information that supports and specifies an existing theoretical position.

Riesch's work exemplifies how an empirical case study, combined with sociological theory, can benefit the philosophical study of scientific practice. By means of qualitative discourse analysis, Riesch studies the question of how reductionism is represented in a sample of popular science literature on sociobiology, evolutionary psychology or Nature/Nuture debates. He shows that reductionism has become an "identity marker" by which popular science authors signal their adherence to a wider social identity. The meaning of 'reductionism' is, thus, to be interpreted according to the stance which the author takes in the debate. Given this fact, Riesch argues that any deeper exploration of philosophical concepts in the thinking of practicing scientists has to take into account sociological factors that shape their interpretation. For a philosophical analysis to state that scientists use concepts in a 'confused', incoherent or inconsequent way is not enough. Instead, it is necessary to understand where and why possible conceptual confusion arises. To do so, Riesch points out, philosophical analyses profit from considering sociological theory.

Addressing a recently growing interest in case studies of scientific explanations in philosophy of science and science education, *Goddiksen* provides an empirical method for collecting and comparing exemplar explanations provided to science students. The aim of his method is to explicate possible qualitative differences between explanations in different disciplinary contexts. The problem that his method addresses is the challenge that an empirical study of explanations needs to identify explanations without presuming the very features the presence, absence or variation of which it seeks to study. To the philosophical eye, it is often not apparent what a scientific explanation is. Drawing on his study of explanations of thermodynamics in physics and chemistry textbooks, Goddiksen discusses various strategies for identifying explanations in science textbooks, such as key word based sampling, and shows what results they can yield.

Jukola examines two cases of commercialized biomedical research in order to discuss the applicability of Helen Longino's view on the objectivity of science to current scientific practice and sketch ways in which Longino's account might need to be extended or specified. The author argues that it is of particular importance for philosophical analyses of scientific objectivity to attend to the extra-scientific influences on research practices. When, e.g., the distribution of private funding threatens the scientific objectivity of research, strictly scientific problems have roots outside science, and in such cases philosophy of science benefits from empirical knowledge about the mechanisms by which extra-scientific influences shape scientific practice.

2.3 Part III: "Empirical Philosophy of Science and HPS"

The volume's last two papers approach the relationship of philosophical theorizing and qualitative empirical insight by revisiting the debates that accompanied philosophy of science's long-standing involvement with the use of historical data, i.e., another kind of qualitative empirical data.

The paper by *Thorén* returns to the claim that history of science cannot provide solutions to the problems that philosophy of science studies. However, even if the claim was true, it would not imply that there cannot be a fruitful relationship between history and philosophy of science. The author argues that the relationship between the two disciplines is best understood as a transfer of problems, and he shows that such problem transfer has established genuine interdisciplinarity between history and philosophy of science. Moreover, he points out how philosophy's appropriation of problems raised in the historical, i.e., empirical, study of science has opened parts of philosophy of science towards empirical knowledge and, thereby, initiated deep changes in its disciplinary understanding.

Allchin addresses the question of how history's descriptive accounts can contribute to an empirically-informed, yet ultimately normative philosophy of science. Allchin explores an approach to this question that is popular among many philosophers of science, an approach adopting abstract philosophical norms about scientific knowledge but remaining uncommitted to the details of scientific practice by which these norms may be achieved. The study of these methods is left to historians of science. As Allchin points out, however, philosophical accounts that articulate the ways in which scientists achieve the normative goals of their inquiry are more complete and applicable to actual scientific practice. The author develops this point by elaborating on the challenge that anomalies pose for philosophical accounts of science, and he illustrates his argument with a case study from cellular biochemistry. A purely abstract philosophy of science, Allchin argues, fails to recognize those strategies by which scientists can solve disagreement upon analogies.

References

Andersen, H., Barker, P., Chen, X.: Kuhn's mature philosophy of science and cognitive science. Philos. Psychol. **9**, 347–363 (1996)
Bailer-Jones, D.M.: Scientists' thoughts on scientific models. Perspect. Sci. **10**(3), 275–301 (2002)
Barker, P., Chen, X., Andersen, H.: Kuhn on concepts and categorization. In: Nickles, T. (ed.) Thomas Kuhn, 212–245. Cambridge University Press, Cambridge (2001)
Bogdan, R., Taylor, S.J.: Introduction to Qualitative Research Methods: A Phenomenological Approach to Social Science. Wiley, New York (1975)
Brinkmann, S., Jacobsen, M.H., Kristiansen, S.: Historical overview of qualitative research in the social sciences. In: Leavy, P. (ed.) The Oxford Handbook of Qualitative Research, 17–42. Oxford University Press, Oxford (2014)
Burian, R.M.: More than a marriage of convenience: on the inextricability of history and philosophy of science. Philos. Sci. **44**, 1–42 (1977)
Burian, R.M.: The dilemma of case studies resolved: the virtues of using case studies in the history and philosophy of science. Perspect. Sci. **9**(4), 383–404 (2001)
Chandrasekharan, S., Nersessian, N.J.: Building cognition: the construction of computational representations for scientific discovery. Cognit. Sci. **33**, 267–272 (2015)
Chang, H.: Beyond case-studies: history as philosophy. In: Mauskopf, S., Schmaltz, T. (eds.) Integrating History and Philosophy of Science, 109–124. Springer, Dordrecht (2012)
Chen, X., Barker, P., Andersen, H.: Kuhn's theory of scientific revolutions and cognitive psychology. Philos. Psychol. **11**, 5–28 (1998)
Chubin, D.E.: The conceptualization of scientific specialties. Sociol. Quart. **17**, 448–476 (1976)
Cole, J.R., Cole, S.: Social Stratification in Science. University of Chicago Press, Chicago (1973)
Darden, L.: Theory Change in Science: Strategies from Mendelian Genetics. Oxford University Press, Oxford (1991)
de Vries, M., van Leeuwen, E.: Reflective equilibrium and empirical data: Third person moral experiences in empirical medical ethics. Bioethics **24**(9), 490–498 (2010)
Denzin, N.K., Lincoln, Y.S.: Introduction: the discipline and practice of qualitative research. In: Denzin, N.K., Lincoln, Y.S. (eds.) Handbook in Qualitative Research, 1–19. Sage: Thousand Oaks (2000)
Feigl, H.: Beyond peaceful coexistence. In: Stuewer, R.H. (ed.) Historical and Philosophical Perspectives of Science, 3–11. University of Minnesota Press, Minneapolis (1970)
Filstead, W.J.: Qualitative Methodology: Firsthand Involvement with the Social World. Markham, Chicago (1970)
Giere, R.N.: History and philosophy of science: intimate relationship or marriage of convenience? British J. Philos. Sci. **24**, 282–297 (1973)
Giere, R.N.: Explaining Science: A Cognitive Approach. University of Chicago Press, Chicago (1988)
Glaser, B.G., Strauss, A.L.: The Discovery of Grounded Theory. Aldine, Chicago (1967)
Goldman, A.I.: Epistemology and Cognition. Harvard University Press, Cambridge (1986)
Gooding, D.: Experiment and the Making of Meaning: Human Agency in Scientific Observation and Experiment. Kluwer, Dordrecht (1990)
Gutting, G.: Michel Foucault's Archaeology of Scientific Reason. Cambridge University Press, Cambridge (1989)

Gutting, G.: Continental Philosophy of Science. Blackwell, Oxford (2005)
Hagstrom, W.O.: The Scientific Community, vol. 304. Basic Books, New York (1965)
Kitcher, P.: The Advancement of Science. Oxford University Press, Oxford (1993)
Knorr-Cetina, K.D.: The Manufacture of Knowledge. Pergman, Oxford (1981)
Kon, A.A.: The Role of Empirical Research in Bioethics. Am. J. Bioeth. **9**, 59–65 (2009)
Latour, B., Woolgar, S.: Laboratory Life: The Construction of Scientific Facts. Princeton University Press, Princeton (1979)
Laudan, L.: Progress and Its Problems: Toward a Theory of Scientific Growth. MIT Press, Cambridge (1977)
Leget, C., Borry, P., De Vries, R.: 'Nobody tosses a dwarf!' The relation between the empirical and the normative reexamined. Bioethics **23**(4), 226–235 (2009)
MacLeod, M., Nersessian, N.J.: Coupling simulation and experiment: the bimodal strategy in integrative systems biology. Stud. Hist. Philos. Biol. Biomed. Sci. **44**(4), 572–584 (2013)
McMullin, E.: The history and philosophy of science: a taxonomy. In: Stuewer, Roger H. (ed.) Historical and Philosophical Perspectives of Science, 12–67. University of Minnesota Press, Minneapolis (1970)
Mansnerus, E.: Understanding and governing public health risks by modeling. In: Handbook of Risk Theory, 213–237. Springer, Netherlands (2012)
Mauskopf, S., Schmaltz, T.: Integrating History and Philosophy of Science. Springer, Dordrecht (2012)
Mattila, Erika: Interdisciplinarity "in the making": modeling infectious diseases. Perspect. Sci. **13**(4), 531–553 (2005)
Merton, R.K.: The Sociology of Science. Theoretical and Empirical Investigations. University of Chicago Press, Chicago (1973)
Nersessian, N.J.: Faraday to Einstein: Constructing Meaning in Scientific Theories. Martinus Nijhoff, Dordrecht (1984)
Nersessian, N.: A cognitive-historical approach to meaning in scientific theories. In: Nersessian, N. (ed.) The Process of Science, 161–177. Martinus Nihjoff, Dordrecht (1987)
Nersessian, N.J.: How do scientists think? Capturing the dynamics of conceptual change in science. In: Ronald N. Giere (ed.) Cognitive Models of Science, 3–44. University of Minnesota Press, Minneapolis (1992)
Nersessian, N.: Opening the black box: cognitive science and history of science. Osiris **10**, 194–214 (1995)
Nersessian, N.J.: The cognitive-cultural systems of the research laboratory. Organiz. Stud. **27**(1), 125–145 (2006)
Nersessian, N.J., Patton, C.: Model-based reasoning in interdisciplinary engineering. In: Meijers, A.W.M. (ed.) The Handbook of the Philosophy of Technology and Engineering Sciences, 678–718. Springer, Berlin (2009)
Nickles, T.: Philosophy of science and history of science. Osiris **10** (2nd series), 138–63 (1995)
Pitt, J.: The dilemma of case studies. Perspect. Sci. **9**(4), 373–382 (2001)
Riesch, H.: Simple or simplistic? Scientists' views on Occam's Razor. Theoria. Revista de Teoría, Historia y Fundamentos de la Ciencia **25**(1), 75–90 (2010)
Quine, V.v.O.: Epistemology naturalized. Ontological Relativity and Other Essays, 69–90. Columbia University Press, New York (1969)
Schickore, J.: More thoughts on HPS: another 20 years later. Perspect. Sci. **19**(4), 453–481 (2011)
Thagard, P.: Computational Philosophy of Science. MIT Press, Cambridge (1988)
Vidich, A.J., Lyman, S.: Qualitative methods: their history in sociology and anthropology. In: Denzin, N.K., Lincoln, Y.S. (eds.) Handbook of Qualitative Research, 23–59. Sage, London (1994)
Wagenknecht, S.: Facing the incompleteness of epistemic trust: Managing dependence in scientific practice. Social Epistemol. **29**(2), 160–184 (2015)
Wagenknecht, S.: Opaque and translucent epistemic dependence in collaborative scientific practice. Episteme **11**(04), 475–492 (2014)
Wouters, P.F.: The citation culture. PhD dissertation, University of Amsterdam, Amsterdam (1999)
Zuckerman, H.: Theory choice and problem choice in science. Sociol. Inquiry **48**(3–4), 65–95 (1978)

Part I
Foundations

Prolegomena to an Empirical Philosophy of Science

Lisa M. Osbeck and Nancy J. Nersessian

Abstract We identify and address a set of foundational questions relevant to the project of an empirical philosophy of science, the most basic of which is the nature of the empirical. We review the task of distinguishing empirical from non-empirical questions by providing examples from our analysis of cognitive and learning practices in biomedical engineering laboratories. We emphasize that the empirical should be understood as rooted in the instrument, and that the instrument comprises the researcher, which includes elusive factors such as disciplinary identity, disposition, and values. The implications of this claim are examined in relation to three empirical approaches to the philosophy of science: historical, qualitative, and experimental.

Keywords Method · Foundations · Epistemic values · Historical analysis · Ethnography · Experimentation

1 Introduction

The title of our paper, though tongue-in-cheek, harkens back to Kant on purpose. The questions we are asking were present in some form in 1783, and it was with the same basic questions Kant was wrestling: How can we understand the empirical? What are its preconditions and limits? How do we move from the empirical to concepts? In essence, what are the grounds of possibility of science (for him, natural science) and social science?

L.M. Osbeck (✉)
University of West Georgia, Carrollton, GA, USA
e-mail: losbeck@westga.edu

N.J. Nersessian
Department of Psychology, Harvard University, 1160 William James Hall,
33 Kirkland St., Cambridge 02138, MA, USA
e-mail: nancyn@cc.gatech.edu

© Springer International Publishing Switzerland 2015
S. Wagenknecht et al. (eds.), *Empirical Philosophy of Science*,
Studies in Applied Philosophy, Epistemology and Rational Ethics 21,
DOI 10.1007/978-3-319-18600-9_2

It is not our intention to definitely answer the questions we pose, but to open a discussion and call for an effort to confront some preliminary questions and problems attending the use of empirical methods in philosophy of science. The closest we will come to providing an answer for the most basic questions concerning the nature of the empirical (and thus the nature and possibilities of an empirical philosophy of science) is this: *The empirical* is rooted in *the instrument* and cannot be understood apart from it. The instrument, of course, consists of relevant technology and established and reliable methods suited to the using the technology to address a question of interest. However, the point we will develop here is that on a more fundamental level, the instrument comprises also *the researcher* who actively selects and analyzes data.

The researcher's central role in science is easier to appreciate in relation to the non-empirical questions that are part of any investigatory project. By emphasizing the central role of the researcher in empirical questions as well, it might be thought that we risk collapsing empirical with non-empirical questions. But on the contrary, the delineation of empirical from non-empirical questions is the most basic issue in play, for science as for an empirical philosophy of science. We do not have a formula for delineation of empirical from non-empirical questions, yet we can provide examples of delineating efforts and the outcomes of these efforts. To do so we draw from the history of efforts at delineation in the discipline of psychology, where the debate has been long and vigorous. As illustrations, we draw upon our own investigation that entails years of collecting and analyzing historical and ethnographic data to address philosophical questions about the nature of science practice in physics and in bioengineering science. After a brief introduction, we provide two examples of questions prerequisite to the empirical study of science that are not themselves empirical questions and two examples of empirical findings that in our case *have* informed our understanding of the sciences we study, which have broader implications for our understanding of science practice in general.

2 Non Empirical Questions in an Empirical Investigation of Science

The first non-empirical problem that confronts us is a hornet's nest of troublesome categories upon which the whole enterprise of an empirical philosophy of science can be said to rest. Among the most difficult is 'empirical' itself, though 'empirical' connects with or is embedded in a cluster of interconnected terms and fuzzy categories such as 'method', 'science', and ultimately, 'reality'.

The difficulties surrounding the meaning of 'empirical' infrequently find their way into discussions of empirical methods in philosophy. This is itself problematic. An example is evident in relation to the recent trend of adopting empirical methods from psychology to inform philosophy, including philosophy of science. This "experimental philosophy", is something of a curiosity, because nowhere has the ambiguity of "empirical" created more problems than in the discipline of psychology. Psychology's fraught history and fragile conceptual edifices should stand as a warning rather than a beacon to philosophers when it comes to adopting appropriate methods for philosophy of science. There is a risk of borrowing psychological methods too hastily to inform philosophical questions while ignoring the more than century long debate over their range, fit, and adequacy. At the same time, there are lessons to be gleaned from the history of psychological science.

One lesson concerns the grounds for adopting a particular empirical method, for choosing one method over another. A prominent view is that the method(s) should be appropriate to the empirical reality, the ontology of what is to be investigated. Thus, for example, as concerns psychology, social or collective processes such as myth-making demand interpretative inquiry; study of differences in individual reaction time requires precise measurement and experimental control. This is, in very rough form, Wundt's view (1901). But differences in method and differentiated units of analysis can arise in relation to the same phenomenon when differing perspectives are taken on the phenomenon. At the end of the 19th century, a scant 20 years after the opening of Wundt's laboratory, Titchener described a division within the science of psychology between its structural and functional aspects, between the concern for the 'plan of arrangement' in the mind's 'mass of tangled processes' and concern with the "system of functions" that enables mind to "do" things for us or equips us to "do" (1899, p. 290). The emphasis on structure entailed a reduction; the emphasis on function, a systems view (Ahn et al. 2006). The difference in perspective or emphasis accompanied differing sets of questions, different methods, and different levels of analysis in relation to the same subject matter—consciousness (James 1890; Titchener 1898). The focus of structural psychology called for controlled, laboratory based experimentation; the focus of functional psychology required analysis of how processes function, what mind does and what it allows people to do, sometimes in the laboratory but often *in the contexts of their natural activity*. Titchener acknowledged the differing emphases to be complementary, as reflecting the structural and functional concerns of the science rather than as attributable to ontological dispute or convention ("turf wars"). Nevertheless, what might have remained a removed recognition of different aspects of the science became a point of contention and social positioning, a "violent controversy" (Boring 1929/1950, p. 314) prompting distinct "schools," "systems" or disciplinary provinces by the early 20th century (Angell 1907). Functional psychology all but disappeared as William James and John Dewey abdicated for philosophy, and John Watson sounded a call for replicable, publically verifiable data and an elimination of consciousness as the focus of psychological science. The differences between structural and functional psychologists reflect not the investigation of different psychological processes (phenomena) but different perspectives on the same psychological process—consciousness.

Differences in emphasis are not limited to the period in which the contours of the new science of psychology were in negotiation. In the latter half of the 20th century, computational and situated approaches to cognition proffered different perspectives on the nature of cognition and the kind of empirical phenomena required to study it, which lead to different starting assumptions and investigative methods, and so to different research programs. The original *physical symbol system* view of cognition (Newell 1980) focused on more professional cognitive tasks such as chess playing and disease diagnosis, while the more recent *situated* perspective (Lave 1988) focused on mundane tasks such as arithmetic use by grocery shoppers and dieters (see, Bredo 1994 for a succinct summary).

2.1 Non-empirical Question #1: What Counts as an Empirical Approach to the Study of Science?

The point we wish to emphasize is that across psychology's history as a formal discipline there has been little agreement about what is to be counted as an empirical approach, included as legitimate data, as the 'facts' of the science, let alone how the facts should be evaluated. Of course, the question of what is to be counted as the empirical reality is informed by the theoretical assumptions in play, by the model from which one is working. But these assumptions and models are themselves influenced by a complex set of factors that include *disposition, identity, and value*.

An emphasis on disposition and value is embedded in Angell's emphasis on function offered in the preface to his textbook on psychology: "Psychologists have hitherto devoted the larger part of their energy to investigating the structure of the mind. Of late, however, there has been manifest a *disposition* to deal more fully with its functional and genetic phases. To determine how consciousness develops and how it operates is *felt* to be *quite as important* as the discovery of its constituent elements" (Angell 1904, p. iii).

The feeling of what is "quite as important" is, we claim, a matter of *epistemic value*. The source of differences in epistemic value is itself a hugely complicated question. In disciplinary practices, such as adopting a particular method or evaluative approach (e.g. a reductionist vs. a systems approach), surely disposition implicates not only a process of socialization to a specific academic community in which that approach is favored but also cognitive style, such that one is more readily drawn to and embraces the values, attitudes, and epistemic assumptions sanctioned within the community of which one becomes a part. That is, value intertwines with academic identity. Likewise, identity has social and personal dimensions, a personal story line and a social history by virtue of the groups with which one actively and passively identifies, along with the history of these groups. These are important aspects of what accounts for the general theoretical models used and stances taken—the scientific perspective (Giere 2006). As is the case in psychology, such differences are a force to reckon with in relation to the emerging empirical philosophy of science.

There are three main approaches to empirical inquiry in philosophy of science that carry differences in identity and epistemic value: historical, observational/ethnographic, and experimental that we examine below. The main point we wish to make, and what we intend by the idea of rooting the empirical in the instrument is this: What is to be counted as empirical in an empirical philosophy of science is hardly itself an empirical question. Rather, what is taken as adequately or appropriately empirical represents a choice and a commitment, either an explicit choice made on grounds that are largely philosophical, or an implicit choice based on disposition and membership in a community that shares a set of epistemic values.

For our purposes, then the empirical question of how epistemic values are formed may not be as important to ask as how they function, what affordances and constraints they offer in relation to inquiry in general and the philosophy of science in particular. In the interest of specificity, we now examine how questions of epistemic values and identity are implicated in relation to the three prominent approaches to empirical inquiry in philosophy of science: historical, and more recent observational/ethnographic, and experimental approaches. We examine the affordances and limitations of these identities and their associated attitudes and values. Each of these empirical approaches is rooted in traditions of analysis that have longer histories and broader scope than just philosophy of science. Note that by implicating values and dispositional factors we bracket ontological considerations relevant to the three approaches, i.e., what is appropriate to the subject matter. Instead, we focus on differences among empirical approaches, and, focusing on the relations between approach, disposition, and value, we specifically note the importance of rooting the 'empirical' in the instrument, i.e., the researcher.

Historical Analysis. Using historical data and methods of analysis has a very long history in empirical philosophy of science, with accounts too numerous to list. The scientific status of historical inquiry and analysis has been in dispute at least since Dilthey (1910/2002), with radically different approaches to historical interpretation emerging in the twentieth century. Despite differences in assumption and approach, the limitations of historical analysis in general are easy to identify. A certain level of vagueness and indeterminacy is acceptable and inescapable. Although there are methods for historical analysis it is simply not possible to codify procedure rigorously, even for historical work deemed positivistic. Hence training or education in historical analysis is distinguished by the substantial role played by apprenticeship. One develops a "feeling for analysis" under the guidance of a mentor.

The absence of a prescribed method in historical analysis offers the advantages of relative autonomy. There is a great deal of flexibility in relation to one's research questions. Affordances include freedom in relation to the selection of cases or episodes for analysis and the kinds of data to examine, and freedom in relation to analytic procedure. One can be more inventive with ones methods. It is of course the degree of freedom involved that has served as the point of contention among different approaches to historical interpretation (e.g. Beiser 2011). Historians have long used the resources, insights, and analytical methods of many other disciplines to deepen historical understanding. Historians of science, too, frequently draw on the resources of other disciplines—anthropology, economics, political science, literature, sociology, cognitive science—to further their analyses. What resources outside of history one draws upon in any given analysis depend on the questions one is asking.

In practical terms, historical analysis can be conducted without the benefit of a research team. This is also a potential limitation: Typically historical analyses are carried out by individuals, even if through mentorship. Although not by any means a necessary component of historical analysis, it is a norm. A historical analysis is typically a single perspective on a data set. Finally, the available data are a fait accompli. There is no opportunity to collect further data to inform a question arising in the middle of analysis. For example, in the absence of any records on the subject, one can only make reasonable conjectures about, for instance, what was the problem that led to an 8-month lag between the first two parts and the third part of Maxwell's 1861-2 analysis of "physical lines of force" (Maxwell 1861-2; Nersessian 1984, 2008). Thus it is safe to say that there are questions that cannot be definitively answered by historical analysis alone.

Qualitative Analysis. By qualitative analysis we primarily refer to analyses based on observational, interview and ethnographic investigations of scientists in real world contexts of practice. This kind of data collection and analysis is relatively recent in philosophy of science. In addition to our own research, recent exemplars are Calvert and Fujimora (2011), Kastenhofer (2013), Knuuttila and Loettgers (2011). The affordances of ethnographic study are in many ways similar to those of historical analysis, but there some important differences. Like historical analysis, ethnographic analysis affords the opportunity to enter into and evaluate the fullness of the life-world, the lived complexities of scientists and the irreducibly rich contexts of their problem solving, thereby avoiding an artificial abstraction away from these complexities.

A good deal of freedom is afforded with qualitative analysis, the nature of which differs in some aspects from that of historical analysis. One can decide on the form of data to collect, what are the sufficient and important data needed to understand the science, rather than having to rely on the data that are left behind. One is "there" in a way that opens opportunities to make spontaneous decisions about what might be interesting and important. There is a much better possibility of collecting the kind of data that suits and informs ones research questions, than is the case with historical data. One can create new data at will with new observations and interviews.

On the other hand, as with historical analysis, one is somewhat at the mercy of the participants (the scientists), what they are willing and able to provide. In an example from our study, in the first lab we investigated we asked to see researchers' laboratory notebooks. We had assumed that laboratory notebooks, as with our experience with historical analysis, were an important part of any experimental/laboratory practice. We thought, in this case in particular, that these would provide us with a record of the development of the physical models currently in use. However, when asked to produce them for our study, the principal investigator asked "what notebooks?" They did indeed keep relevant information about specific experiments in documents on their computers, but these were largely strings of numbers devoid of any comments or reflection. Thus there is still the problem that scientists engaged in real world contexts of practice might not offer the kinds of data we feel are important. There are constraints on top down analysis, that is, because the data might not be available to answer the questions we have.

In terms of training, one can dabble a bit with qualitative analysis just as one can with historical analysis, to a degree that is not possible with experimental or quantitative approaches. That is, much of the learning is "on the ground," honed through apprenticeship of various forms. One can get ones feet wet in qualitative inquiry and analysis without a great deal of detailed preparation. There is not a rigorous cannon of procedure which one should master before becoming involved in a research project. Proponents of qualitative analysis, however, unlike historical, have devoted considerable attention to methods. Here we focus on debates within psychology, but the issues have been raised in sociology and anthropology as well. Among psychologists, the use of qualitative and interpretive methods have incited controversy within a science designed to achieve independence from philosophy by means of positivistic methods to address questions about mind. Various forms of naturalistic inquiry and exploration of experience were met with criticism from the very beginning, in large part because they were regarded as problematically importing philosophical assumptions into psychological inquiry (as if controlled laboratory experimentation did not do so!): "(A)nything approaching a complete and permanent divorce of psychology from philosophy is surely improbable so long as one cultivates the functionalist faith" (Angell 1907, p. 90).

Challenges to the legitimacy of qualitative analysis as a foundation for knowledge and questions concerning their generalizability and predictive utility have accompanied their use historically. Of late there have been increasing efforts to name and describe various systems of qualitative coding and analysis. Qualitative methods books are proliferating, as are systematic attempts to distinguish the different approaches from one another and occasionally to analyze their common fundaments (e.g., Wertz et al. 2011, is a recent example of this effort). We suspect that psychology's continuing obsession with method has had a great deal to do with the emphasis on distinguishing qualitative approaches and attending to their unique forms of rigor. Part of the justification of procedure comes from its belonging to a recognizable and named category of procedure, despite the fact that new research contexts and questions might call for innovations in procedure. Concerns with establishing inter-rater reliability in developing codes and similar matters reflect the same trend. By contrast, anthropology has not had to justify its methods in the same way. The focus has been on the researcher as instrument, as tool. It is enough that the researcher is "there," in the setting in which the inquiry takes place; there is implicit trust in the veracity of the observations of an eyewitness. Traditionally, ethnographic research is carried out by an individual, although within philosophy of science (including our own research) there has been a trend towards what could be called "team ethnography," in which multiple perspectives of several researchers are brought to bear on an interpretation.

Among the reasons for concern with legitimacy is that much of what passes for procedure entails seemingly irreducible acts of insight; thus much is not amenable to "neutral" description, let alone replication. Qualitative analyses cannot pass the kinds of reliability tests established for the purpose of evaluating quantitative data, prompting charges that qualitative analysis represents "mere storytelling." In addition to requiring a willingness to engage methods that remain controversial in some corners, qualitative analysis, like historical analysis, suits some dispositions better than others. It requires the ability to abstract from reams of data stemming from various sources to form insights, make and hold tenuous connections, generate possibilities. One must have a high degree of tolerance for ambiguity and uncertainty, must be comfortable with messiness and feeling out of control, leaving things open, being surprised, not knowing where one is going. Qualitative analysis does, however, offer the possibility of collaboration and bringing multiple perspectives to bear on an interpretation. Our interdisciplinary research group has had considerable, fruitful experience with perspective sharing or exchange. All of what we are calling the affordances of qualitative analysis are, from another perspective, intractable limitations. One relinquishes control, precision of description, predictive ability.

Experimental/Correlational/Descriptive Statistical Analysis. That which is lost with qualitative analysis might be gained with explicitly quantitative approaches, and even more so approaches that incorporate controlled experimentation to inform scientific reasoning. The recent development of an "experimental philosophy" has largely been confined to the philosophy of mind. Contemporary work in experimental philosophy harnesses the methods of social science, especially psychology, to investigate and challenge prevailing assumptions in the context of philosophy of mind (Deutsch 2009; Knobe and Nichols 2013; Machery et al. 2004). Within the methodological traditions from which experimental philosophers draw, the most robust efforts have all of the weight of experimental logic behind them, providing a grounding for inferences that can never be matched by historical and qualitative approaches. Abstraction, precision, control, detail, and statistical power are formidable allies. The corresponding dispositional qualities are not difficult to identify. Not surprisingly, precision comes at a price. Vagueness is not acceptable. In psychology, at least, the training required to do this kind of analysis well is extensive; it requires rigorous training in experimental design and statistical analysis. One cannot dabble in it. It is not enough just to learn some statistics. A significant limitation, though, is that the range of questions that can be asked is much narrower in scope. One must be comfortable with a restricted range of questions and possibilities for addressing them.

Summary. Beyond training considerations, the important point is that all three approaches lend themselves to different *dispositions*. They should be regarded as complementary, not in competition, because they are aimed at different questions or different levels of question, which the observational and analytic powers of different researchers equip them to address differentially.

Of course, in suggesting that different empirical approaches suit different dispositional qualities, we risk reducing epistemology to psychology. That is not our aim. We aim only to point out that in practice it is not *merely* ontological considerations that determine empirical approach. But of course ontological considerations should play the major role. Thus we return now to the question of ontology as relevant to the philosophy of science, namely the unit of analysis chosen for the investigation of science practice.

2.2 Non-empirical Question #2: What Is the Appropriate Unit of Analysis for an Empirical Investigation of Science?

The question of the unit of analysis appropriate to an investigation is not an empirical one. That is, it is not empirical apart from the sense in which history provides a guide to decisions that have been made in previous efforts to investigate phenomena of the kind in question. The considerations are here are ontological, because the unit of analysis concerns the nature of the object under investigation in philosophy of science. Unit of analysis determines perspective: it involves a decision about where to look, how widely to extend the gaze. Decisions about where and how widely to look, in turn, implicate a set of constraints on what can be "seen," along with the aspects of the phenomenon to which one will be "blind" (Giere 2006). Decisions about where to look, or what to look at, are then followed by decisions about how best to organize analysis in relation to the level of complexity of the subject matter. In short, the epistemological considerations follow from the choice of unit of analysis. The choice of unit of analysis can be influenced by many things. It is the cannon of good science that methods should follow from, not lead to one's questions. The unit of analysis, however, may be co-implicated in a question, may lead to a question, or may follow from it.

In our own work, our choice of unit of analysis follows from our problem formulation. We have identified what we have called an "integration problem" in science studies (and in psychology, for that matter). The majority of cognitive studies of science have proceeded in relative isolation from social and cultural studies of science, while the latter have largely ignored the need to address cognitive dimensions Both Longino (2002) and Nersessian (2005) separately have pointed to the implicit acceptance of a rational–social dichotomy in both philosophy of science and science studies. There are conceptual problems with any such dichotomy, as Vygotsky (1978) and scores of others have made clear. Therefore the unit of analysis for an adequate understanding of science must be one that does not perpetuate such a divide. For our purposes, we have found it very useful to select as our unit of analysis the *acting person*. The acting person is a social, cultural, and cognitive being with a particular experience, disposition, and identity.

What is implied in ascribing the label of "person" is a longstanding problem with a great deal of baggage, usually relating to intentionality, rationality, language-use, rule-following, or individuality/particularity, depending on the context and purpose. The choice of person as unit of analysis for the study of science may seem peculiar. Michael Polanyi acknowledged the seeming tension, even contradiction, between 'persons' or 'the personal' and science in his preface to *Personal Knowledge* (1958), noting the impersonal and universal features typically emphasized in relation to science and assumed to be necessary to a proper understanding of its authoritative grounding. In turn, 'the personal' is associated with variation, deviation, difference, contamination (Titchener 1912).

However, if our focus turns to the empirical dimensions, to science as it is actually practiced in real world settings, rather than as an idealized conception of methods, logic, and products, attention to the particularity and experience of the person, the scientist, is a necessary complement to social, historical, and cognitive dimensions of analysis. Thomas Kuhn appears to have arrived at something like this insight:

Just because the emergence of a new theory breaks with one tradition of scientific practice and introduces a new one conducted under different rules and within a different universe of discourse, it is likely to occur only when the first tradition is felt to have gone badly astray. That remark is, however, no more than a prelude to the investigation of the crisis-state, and, unfortunately, the questions to which it leads demand the competence of the psychologist even more than that of the historian. What is extraordinary research like? How is anomaly made law-like? *How do scientists proceed when aware only that something has gone fundamentally wrong at a level with which their training has not equipped them to deal?* Those questions need far more investigation, and *it ought not to be all historical.* (1962, pp. 85–86, emphasis added)

Kuhn's remarks point to the need for enhanced understanding of the overall function of personal factors in the hows and whys of scientific practice—such as how a scientist's awareness of her own shortcomings in relation to a new direction might influence her readiness or resistance to change. This is a question of learning history and identity, of positioning, and broadly speaking, perspective. There are also implications of emotional involvement. Adequate characterization of science practice must at some point come to terms with the problem of the personal, with the fact that people bring different levels of cognitive ability, different interests, goals, desires, problems, experiences and collaborative relationships into any laboratory, no matter how systematic its proceedings. There has been insufficient effort to carefully theorize how these differences impact the "organized, artful practices" that constitute rational achievements in real world settings (Garfinkel 1967, p. 34). A variation on this question is whether "the personal" dimension might be understood not merely as a source of impurity or impediment but as a set of processes that enhance and indeed, enable science.

We have argued that emphasis on the act*ing* person encompasses both the intentional quality of action and the social meaning or force of acts accomplished through the actions, for the intentional performances of persons (actions) always take place within socially negotiated or inherited contexts of social meaning (Osbeck et al. 2011). An understanding of scientific practices as normatively structured by sanctioned methods, communal ideals, and field-specific projects does not alter the fact that science consists in activities of persons, nor even does the recognition that economic and political controls are driving scientific agendas in a broad scale way. Persons act to collaborate with other persons using the tools available to them, always in relation to goals, desires, aspirations, and values both collective (values held by the scientific community at large, such as advancing knowledge) and particular (advancing one's career, solving a problem, obtaining closure). The acting person as an analytic unit then integrates intentionality, creativity, and social normativity; it represents an inherently integrated focus of analysis.

Given our focus on the acting person, one promising direction is to concentrate more intently on ways in which identity is implicated in scientific reasoning. The utility of identity for the present purposes lies in the fact that the category historically has involved both personal and social dimensions, an experience of one's own unique history, place, aspirations, and meanings and the groups or social formations to which one claims belonging, prompting the rather clumsy distinction between personal and social identity (Turner 1982). More strongly, relational identity has been suggested as a precondition for the experience of personal identity (e.g. Mead 1934).

Identity is a notoriously ambiguous category, but it implicates a constellation of concepts important for understanding science: value(s), emotion, embodiment, the anticipated and experienced gaze of the other. It is a form of enactment despite the experience of continuity and permanence. The close relation of identity to social positioning means that identities can be seen to establish the possibilities of action. They have epistemic effects, are integrally related to problem solving, influencing what and how one feels able or entitled or do within the wide range of practices that constitute science (see Osbeck et al. 2011, Chap. 5). Such considerations are increasingly important with the growing trends towards interdisciplinary and transdisciplinary collaborations in science.

3 Empirical Questions in an Empirical Investigation of Science

The purpose of our remarks so far has been to identify aspects of our analysis that constitute non-empirical questions in the empirical investigation of science. We turn now in the other direction, to give examples from our own practice that *have* been directly informed by empirical investigation of science practice. We draw on two empirical investigations. The first is the multi-year cognitive-historical study Nersessian conducted of the formation of the electromagnetic field concept. Interpreting the historical data leading to the development of various electromagnetic field concepts from the mid-1800s to the early 1900s required her to develop a reflexive method of analysis (Nersessian 1984, 1995). "Cognitive-historical analysis" examines historical records in light of cognitive science investigations into mundane reasoning and representation and feeds back the analysis of scientific cognitive practices into the development of cognitive theory. The second is the multi-year ethnographic study we have been conducting of bioengineering sciences laboratories and, more recently, integrative systems biology labs. Our attention is both to the cultural organization of each laboratory setting and the participatory stance of each researcher in relation to biological phenomena, cognitive tools (e.g. models) and instrumentation central to the science. We regard cognitive processes as *system phenomena*, that is, as distributed across persons and artifacts and situated in physical and cultural contexts (e.g. Hutchins 1995a, b; Greeno 1998; Clark 2003; Nersessian et al. 2003) with cognitive activities made possible (afforded) or constrained by the specific properties and composition of environments in which reasoning takes place. We have elsewhere described the laboratory as a cognitive-cultural system in that cognition and culture are co-implicated (Nersessian 2005).

3.1 Model Based Reasoning

When Nersessian began investigating the archival records of Faraday, in particular his Diary, she was struck by the abundance of sketches along the margins and elsewhere that seemed to be playing some role in his reasoning about the phenomena he was investigating at that time. Up to that point she had been indoctrinated as a scientist and then as a philosopher of science with the idea that scientific inference is inductive or hypothetico-deductive reasoning over propositional representations. Although her science teachers sometimes drew diagrams, they never discussed why this might be important. There was virtually no discussion of visual representations in the philosophical literature, and what there was pointed to their role as "mere aids" to reasoning (which is logic-based). The historical literature had likewise tended to ignore them, but just at that time some accounts emerged that looked primarily at the communicative role they serve (e.g. Rudwick 1985). Faraday however appeared to be reasoning through or by means of his sketches and she could tie the articulation of his concept of field directly to specific visual representations. Maxwell, too, seemed to be reasoning by means of a visual representation of what he called a "physical analogy" (Maxwell 1861-2) and with this and other diagrams in his papers, he gave instructions for how the observer should simulate motion of the elements of the diagrammatic representation their imagination. He also wrote several accounts on the importance of physical analogies as a method of discovery. However, his analogies were noted explicitly in both the philosophical and historical literatures as "merely suggestive (Heimann 1970), of "slight" heuristic value" (Chalmers 1973) and at worst as post hoc fabrications, while "the results were known to him by other means" (Duhem 1902). The exception was Hesse (1963) who tried to develop an account of analogy and discussed Maxwell, but was curiously silent about the 1861-2 paper where the physical analogy seemed to be playing a generative role in Maxwell's initial formulation of the field equations. To make a long story short, Nersessian began to think these data should not be considered ancillary, but that these visual representations, analogies, and thought experiments (prevalent in the records of the practices of other historical scientists as well) constituted a form of creative reasoning—what she called "model-based reasoning." It took another 20 years of philosophical, historical, and cognitive science research to articulate the nature of model-based reasoning, including its cognitive basis and how it produces conceptual innovations (Nersessian 1992, 2002, 2008). Expanding from the insights deriving from historical data, our bioengineering laboratory studies over the past 12 years have been looking into the creative roles of model-based reasoning more broadly than conceptual innovation, now focusing on physical and computational models and simulations.

3.2 Relation of Emotion to Problem Solving

We began our investigation of bioengineering laboratories with the explicit goal of characterizing the nature of the cognitive and socio-cultural practices exhibited by researchers across each lab. While coding interview text along these lines we were struck by a good many passages with a decidedly affective tone. Others seemed clearly expressive of desires, goals, and aspirations. We began to assign codes for affective and motivational content and found that it quite naturally sorted itself into three categories of expression: (1) overt expressions of excitement and frustration; (2) metaphorical and figurative expressions in scientists' descriptions of practice; and (3) anthropomorphisms involving an attribution of emotional states to objects, artifacts, and devices. We described these as three classes of affective expression. The important overall point of this analysis is that it demonstrated how closely intertwined affective expression and problem-solving efforts seem to be. We realized how our data implicated ways emotion figures into cognitive acts, that one cannot be entirely disentangled from them without considerable abstraction away from the real world phenomenon of science practice. In turn, we were able to analyze the functional significance of emotion in the overall situation of the laboratory, of which the anthropomorphic expressions are most interesting and significant. In brief, the functional benefits of anthropomorphism are of two related kinds: First, the attribution of emotional states through anthropomorphism reflects implicit emotional processes that contribute to the motivation, interest, and attention of the researcher in relation to the objects and entities central to the laboratory's research projects (Osbeck and Nersessian 2013). Secondly, the attribution of emotion carries attributions of agency. That is, objects central to the practice of the scientist are imbued with agency (functionally so) through anthropomorphism, such that they are transformed into working partners with the research scientist in cognitive practices toward shared and individual problem solving goals. We have construed this process of transforming objects into "partners" in problem-solving practices as "cognitive partnering" (Nersessian et al. 2003; Osbeck and Nersessian 2006). Of course the emotional expressions in the interview text, including anthropomorphisms, did not speak for themselves; they required interpretation and analysis. The point is that these insights concerning scientific reasoning would not have been possible in the absence of the empirical analysis. They would not have occurred to us.

Summary. In this section we provided two examples of questions concerning the nature of scientific reasoning that were informed explicitly by empirical research. We first discussed an example from historical analysis, namely Nersessian's discovery that Faraday and Maxwell appeared to be "reasoning through" models of various forms; that these model-building and manipulation processes were integral to their most important discoveries. Secondly, in an ethnographic study of bioengineering science, we discovered that as researchers frequently and quite consistently use anthropomorphic expressions when referring to the physical and computational models that are central to their problem-solving, providing a connection between affective and inferential processes. Although in each case the findings did not interpret themselves, we were sufficiently "surprised" by them to consider them to be matters of discovery.

We turn now to a more detailed discussion of our current empirical project to provide a context for thinking about the dynamic interplay of empirical and non-empirical questions.

4 Empirical Philosophy of Science in Practice

For illustrative purposes we briefly describe our own approach to empirical analysis that has been informing our investigations of five research laboratories in the bio-engineering science over the past 12 years in order to exemplify the rich affordances of empirical methods for informing philosophical questions about science practice, in our case interdisciplinary science. We draw from the part of our investigation that was situated in two biomedical engineering (BME) research laboratories located on the campus of a major research university in an urban setting. Biomedical engineering may be characterized as an *interdiscipline*, meaning that "melding of knowledge and practices from more than one discipline occurs continually, and significantly new ways of thinking and working are emerging" (Nersessian 2006, p. 127). The labs merge resources from both biology and engineering in the form of researchers, concepts, materials, and methods. In addition to blending academic domains, the labs tend to attract persons with diverse interdisciplinary interests and experiences.

Our study of these interdisciplinary laboratories has also drawn a diverse interdisciplinary team comprising researchers from cognitive science, philosophy of science, psychology, psychoanalysis, linguistics, history of science, and computer science, to understand the learning, reasoning, and problem-solving practices. It has been challenging to draw from these varied influences in such a way that represents adequately disciplinary and dispositional differences while achieving a unified 'voice' for our analysis. We have reconciled these difficulties through regular weekly meetings at which we compare observations and compare and develop interpretations of interviews. Further, our interpretive codes were developed in dyads and refined in the larger group, ensuring that no one interpretive perspective was put forward; rather, we aimed for an integrative perspective.

Our investigation began with the framing assumption that the cognitive practices of each laboratory are both *situated* in the laboratory and *distributed* across systems of interacting persons, artifacts, instruments, and traditions. The situated approach to cognition construes the features of intelligent behavior as arising within and depending upon the constraints and affordances of particular settings, in contrast with a view of cognition as a context-independent abstract set of functions. We understand the laboratory as the physical space, its artifacts, the instruments and devices used for investigation, including technologies specially designed for these purposes, and also as an organized social group that shares an agenda that is to some extent collective. The broader collective agenda underlies and supports the problem-solving goals and strategies of any single researcher at any given time.

The principal investigator of each lab is most obviously involved in setting the collective agenda; however, our analysis has shown that the agenda is *dynamically* influenced by contributions from all members of the laboratory community. We thus construe the laboratory as an "evolving distributed problem-space"—comprising researchers, artifacts, and practices—with permeable boundaries, in that it enables researchers to move between its physical boundaries and the wider community to which the work is connected (Nersessian et al. 2003). Researchers in both labs actively seek new ideas and applications at the cutting edge or frontier of knowledge in their respective fields. They are therefore *creative* environments, which in previous work, Nersessian (2006, 2012) has characterized as distinguishing the study of these laboratories from other problem-solving environments in which the goal is not novelty but precision, such as Hutchins' studies of navigation processes undertaken in landing a plane or steering a ship to harbor (Hutchens 1995a, b). The laboratories are *evolving* systems, with problems, goals, methods, and technologies transforming in response to the activities of its researcher-learners, the entry of new researchers and the departure of others, and with outside collaborations.

Central to the cognitive practices in both laboratories is what we have labeled *traversing the in vivo–in vitro divide*. Research in biomedical engineering must devise ways to emulate selected aspects of in vivo phenomena to a degree of accuracy sufficient to warrant (to the extent possible) transfer of simulation outcomes to the in vivo phenomena. As a result, researchers in both labs design, build, and experiment with hybrid physical in vitro simulation models composed of both living and engineered materials that selectively instantiate what the researchers deem significant features of in vivo systems. Experimentation with these models requires bringing biological and engineering practices together in an investigation into a "multifaceted modeling system" (A-10).

A more detailed description of the purposes and practices of each lab will help to situate our approach to analysis.

4.1 Lab A

Lab A is a tissue engineering laboratory that dates to 1987. During the period of our investigation the overarching research problems were to understand the mechanical dimensions of cell biology, such as in the behavior of endothelial cells in response to shear forces, and to engineer living substitute blood vessels for implantation in the human cardiovascular system. The dual objectives of this lab explicate further the notion of an engineering scientist as having both traditional engineering and basic scientific research goals. Examples of intermediate problems that contributed to the daily work during our investigation included designing and building living tissue —"constructs"—that mimics properties of natural vessels; creating endothelial cells (highly immune sensitive) from adult stem cells and progenitor cells; designing and building environments for mechanically conditioning constructs; and designing means for testing the construct's mechanical strength.

During our study, the main members included a director, one laboratory manager, one postdoctoral researcher, seven PhD graduate students (three graduated while we were there, and the other four graduated after we concluded formal data collection), two MS graduate students, and four long-term undergraduates (two semesters or more). Of the graduate students, two were male and seven were female; the postdoctoral researcher was female. Additional undergraduates from around the country participated in summer internships, and international graduate students and postdocs visited for short periods. Usually the graduate student researchers work on individual projects, often with assistance from undergraduates.

4.2 Lab D

Lab D is a neural engineering laboratory. During the period of our research the overarching research problems were to understand the mechanisms through which neurons "learn" in the brain and, potentially, to use this knowledge to develop aids for neurological deficits. The assumption that guides research in Lab D is that advancing understanding of the mechanisms of learning requires investigating the network properties of neurons. Examples of intermediate problems that contributed to the daily work included developing ways to culture, stimulate, control, record, and image cultured "dishes" of living neuron arrays; designing and constructing feedback environments (robotic and simulated) in which the dish of cultured neurons could "learn;" and using electrophysiology and optical imaging to study plasticity.

During our study the main members included a director, a laboratory manager, a postdoctoral researcher, four PhD graduate students in residence (one left after two years, and three graduated after we concluded formal data collection), a PhD student at another institution who periodically visited and was available via video link, one MS student, six long-term undergraduates, and one volunteer for nearly two years, who was not pursuing a degree (already had a BS) but who helped out with breeding mice. Of the graduate students, two were female and three were male; the postdoc was male. The backgrounds of the researchers in Lab D were more diverse than those in Lab A and included mechanical engineering, electrical engineering, physics, life sciences, chemistry, and microbiology; some were currently students in a BME program. As an institution, the neural engineering laboratory had been in existence for only a few months and was still very much in the process of formation when we began data collection. Because all the projects centered around the "dish" of living neuron here was significantly more interaction among research projects than we witnessed in Lab A. Unlike the traditional independent configuration of Lab A, Lab D is embedded in an open space that is shared by seven faculty members and their postdoctoral researchers, as well as graduate and undergraduate students.

4.3 Data Collection

As noted above, qualitative approaches to inquiry are proliferating in the social sciences. The sheer variety and alleged differences among approaches (e.g. grounded theory, discourse analysis, narrative analysis, phenomenological inquiry) can be daunting. But the basic issues concern the question, problem, or analytic focus, which then have implications for the decision about the particular method most appropriate to use.

For our purposes, an analytic focus on *the acting person,* the scientist, or more specifically *the acting person in normatively structured contexts of practice* (the science laboratory), is an inherently integrated focus. It thus invites an analysis not, e.g., on neural mechanisms in the brain but on acts of coordination or coordinated practices across persons and artifacts occurring in the context of the biomedical engineering research laboratory. Coordinated achievements occur and are demonstrated in both the interviews (conversations) with research scientists and through the practices which are described in detailed field notes on our observations.

Individual Interviews. The question might well be raised why we focus on *interview text* rather than video recordings of laboratory practices. In the learning sciences, video recordings are often used to provide grounds for analysis of complex interactions of persons with one another and with the objects of their practices; enabling consideration of the interrelations of verbal utterances (talk), gestures, use of tools and artifacts, and both routine and novel practices (Jordan and Henderson 1995). We audio and video recorded interactions, but have focused most of our attention on analysis of interview data. It was not possible to record research activities in the labs.

Additionally, we worry about the possibilities of eliminating the affective, motivational, and cognitive particularity of contributors to the collective practice of knowledge construction through accounts that focus solely on interaction. We have no easy solution to the problem of adequately understanding the contribution of the particular to the collective without resorting to an individualistic framework, but the inclusion of the personal dimension of science is necessary to any effort to move beyond the artificial separation of the social and cognitive realms that has dominated accounts of science to date.

Moreover, the use of interviews is a methodological implication that follows from the acting person as an analytic focus. The study of persons should include treating them as persons, which entails enabling them to give reasons, to provide accounts of their activities (Parfit 1984). Scientists do not speak of their subjective or personal investments in their formal reports; research is described as if subjective effects have been eliminated. Yet when scientists discuss their own practices more informally, including in the context of an interview, they include highly personal accounts of their aspirations, influences, accomplishments and failures. That is, the personal dimension emerges as critical to their theoretical commitments and discoveries. Thus, although study of persons in science may well include observation and analysis of their conversational exchanges, it seems also to require talking to them, enabling them to give *reasons* for their specific activities and describe what their practice means to them, to *account for* their practices and research interests. Of course we do not suggest that interviews provide us with an x-ray of our participants' inner world, and an account of what occurs in practice must be compared with ethnographic observations. However, the account provided by an interview (especially a "situated interview" that takes place with the environment—the laboratory—in which the cognitive activities of interest occur), is essential for analyzing the personal contributions of the scientist to the research process.

Through the use of individual interviews with researchers with different levels of expertise, from different disciplinary backgrounds, and at different phases of research, we are able to analyze how the particular learning history, relational networks, affective style, sources of motivation, and epistemic values contribute to what takes place in their own research trajectories and in the relational dynamics of the interdisciplinary space. The interview provides insights into how each scientist understands her work, what it means to her, and how she experiences it. These aspects have tended to be excluded from analysis of science practice to date. Moreover, following Rouse (1996), we regard the interview conversation as part of the wider conversation of science. That is, the felt demand to clarify and explicate their problem solving to a novice outside of their field has been described by some of our participants as contributing to new ways of understanding what they are doing for themselves. Directors of both labs reported that they found that our interviews of their researchers made them more reflective about their practices.

Field Observations. Several members of our group became participant observers of the day-to-day practices in each lab. Each ethnographer "hung out" in a lab, observing and having informal conversations, and attended official laboratory functions (meetings, presentations, dissertation defenses). We estimate that the total time spent in observation of these two labs across our research team is over 800 hours. Team members took field notes on their observations, audiotaped interviews, and video- and audiotaped research meetings (full transcriptions have been completed for 148 interviews and 40 research meetings). We used fieldnotes from the observations to compare with interview data to arrive at our interpretations.

Coordination of Field and Interview Data. Our interdisciplinary investigatory team held regular weekly meetings that allowed us to compare interview data with field notes. We developed and refined coding categories during these meetings. Naturally, the changing composition of the team affected both the style of working together and the specific categories that emerged or received emphasis. Codes that emerged through grounded theory analysis (described later) were "tested" for their applicability and conceptual fit with data recorded as field notes and with a sample of additional interviews. In coordinating interview and field observation data we view ourselves as analytic instruments, relying on the basic human capacities of insight as we engage with the accounts of our participants. Through our dyadic and group coding and refinement of codes we hone these insights by considering multiple perspectives and engaging in discussion, even argument.

4.4 Data Analysis

Development of Codes. We used a coding procedure informed by Grounded Theory (Corbin and Strauss 2008) inasmuch as we attempted to approach the data openly, not looking to confirm the presence of theoretical categories we held to be salient before our research began. Of course, the extent to which we are influenced by our pre-existing theoretical commitments is an open question; the point is that we did not deliberately seek to identify particular themes in the data. We were guided by our research questions but left ourselves open to surprises.

We began by coding a subset of interviews selected to represent different research problems, disciplinary backgrounds, and levels of expertise. Each selected interview was examined line by line, from beginning to end, with the intent of providing a descriptive level for most passages. Tentative codes developed were discussed in larger group meetings. We held detailed discussion about the possible significance and alternative interpretations of the text.

We then grouped codes together under headings that seemed to capture as much as possible their important main theme. For example, model-based understanding and model-based reasoning were included along with model based-description or explanation, which seemed to express situations in which the model was invoked principally for the purposes of explaining a concept to the interviewer. Codes that did not fit easily into one of the main headings were analyzed further for possible overlooked meanings or their fit with other categories. We repeated the process until we could draw no further important distinctions. We then developed and revised a written description of main code categories, with examples of text passages assigned to each category. Main categories, descriptions, and examples were brought to the main research team for feedback and were revisited and in some cases revised after the feedback was received. An Exemplar of the highest level codes that emerged is *Seeking Coherence (sense-making),* which includes subcategories of modeling, framing, positioning, and offering narrative (lab history and personal history).

Case Study and Cognitive-Historical Analysis. In addition to sampling interviews across researchers, another strategy was to focus coding and analysis on interviews with one particular lab member over time, analyzing chronologically one researcher's developmental trajectory from a point very soon after she first entered the laboratory. We used a coding system similar to that used for the analysis across interviews.

Finally, we made use of also of the cognitive-historical method to determine how the representational, methodological, and reasoning practices have been developed and used by researchers in the BME laboratories. Cognitive-historical analysis involves tracking the human and technological contributors to a cognitive system on multiple levels, including their physical shaping and reshaping in response to problems, their changing contributions to the devices developed in the lab and the wider community, and the nature of the concepts that are central to the practice at hand. As with other cognitive-historical analyses, we used a variety and range of historical records over time spans of varying length, ranging from shorter spans defined by the activity itself to spans of decades or more. Although historical in perspective, the focus is on facilitating an understanding of cognition, as well as developing an historical interpretation (Nersessian 1992, 1995, 2008). For this dimension of our study, we collected the publications, grant proposals, dissertation proposals, PowerPoint presentations, laboratory notebooks, emails, materials related to technological artifacts, and interviews on lab history.

4.5 Rigor and Accountability

Although we fully embraced the idea of putting our faith in the instrument of analysis (the researcher) by trusting the interpretations made, we were equally concerned about rigor and holding ourselves accountable for the match or fit between data and interpretation. For instance, we attempted to maximize coding rigor in three ways or phases:

Collaborative Coding. Coding initially took place between two or more members of our research team, ensuring that codes reflected interpretations that seemed plausible to at least two people, usually with different disciplinary backgrounds. Where possible, one of the coders was a person with more advanced knowledge of biosciences to provide help in interpreting specifics of the science.

Group Code Refinement. Updates on coding were presented at the research team's regular weekly meetings, in the context of discussions that occasionally became heated arguments. However, codes were only retained when they seemed plausible and accurate to all team members present. Other codes were adjusted or abandoned to reflect group feedback.

External Audit. After the initial coding scheme was developed, we enlisted an external auditor to review codes and to check them against a data sample. The auditor had expertise with qualitative methods of analysis but was not involved with the project except as an auditor. Thus he had no vested interest in the study's outcome. We provided the auditor with a sample data (interviews), a description of our procedure, and our initial coding scheme (higher and lower order categories). He met with us and provided very favorable feedback on our procedure and interpretations.

Overall, to ensure the "trustworthiness" (Lincoln and Guba 1985) of the findings, we followed Eisner's (2003) three principles: structural corroboration, referential adequacy, and consensual validation. Structural corroboration requires that a sufficient number of data points converge on a conclusion to support the arrived at interpretation. This principle calls for triangulation among different data types, in our case, interviews, field notes, lab meetings and documents. Referential adequacy addresses the richness of the description and interpretation and whether it aligns with member understanding of the same phenomena. It is important to clearly and succinctly explain the properties of each coding category for the sake of transparency. And finally consensual validation refers to the level of inter-rater agreement that can be reached among two or more team members using the coding schemes (as discussed above). Failure to achieve such validation means that the coding scheme is not well corroborated in the data or adequately described. To further ensure the trustworthiness of our findings, a methods consultant advised on procedures for collection and analysis of qualitative data, including interview format, coding procedures, and synthesis of coded material.

5 General Conclusion: Rooting the Empirical in the Instrument

We have attempted to provide some conceptual grounding relevant to the project of an empirically informed philosophy of science. We identified questions important to this grounding, and although we did not attempt to answer them definitively, we provided a guiding framework for understanding the complexities involved. We focused principally on the delineation of empirical from non-empirical questions in an empirical philosophy of science. Although we found ourselves unable to supply a formula for such delineation, we were able to provide examples of what we consider empirical and non-empirical questions in our own work, that help to inform the question of how best to understand "the empirical" in an empirical philosophy of science. Our examples and reflections on both empirical and non-empirical aspects of an empirical philosophy of science point to the same conclusion, namely that we must root our understanding of the empirical "in the instrument." By this we mean to emphasize especially that at the deepest level the instrument comprises the one who engages in the collection and analysis of data. We commented on ways that differences in value and identity, even disposition or temperament (personality), interrelate to the epistemic demands and affordances of three empirical approaches: historical, qualitative (e.g. ethnographic), and experimental analysis. We have tried to make clear that all forms of empirical analysis require *reliable* instruments, including persons who can be trusted to collect adequate data and to analyze it with insight and integrity. We suggest the acting person as a unit of analysis not only as the focus of our investigation of science but as the instrument of empirical philosophy of science regardless of methodological approach.

References

Ahn, A.C., Tewari, M., Poon, C.-S., Phillips, R.S.: The limits of reductionism in medicine: could systems biology offer an alternative? PLoS Med. **3**(6), e208 (2006)
Angell, J.R.: Psychology: An Introductory Study of the Structure and Function of Human Consciousness. Henry Holt, New York (1904)
Angell, J.R.: The Province of Functional Psychology. Psychol. Rev. **14**, 61–91 (1907)
Beiser, F.C.: The German Historicist Tradition. Oxford University Press, Oxford (2011)

Boring, E.G.: A History of Experimental Psychology. Appleton Century Crofts, New York (1950. Original work published 1929)

Bredo, E.: Reconstructing educational psychology: situated cognition and Deweyian pragmatism. Educ. Psychol. **29**(1), 23–35 (1994)

Calvert, J., Fujimura, J.H.: Calculating life? Duelling discourses in interdisciplinary systems biology. Stud. Hist. Philos. Sci. Part C Stud. Hist. Philos. Biol. Biomed. Sci. **42**, 155–163 (2011)

Chalmers, A.F.: Maxwell's methodology and his application of it to electromagnetism. Stud. Hist. Philos. Sci. **4**(2), 107–164 (1973)

Clark, A.: Natural Born Cyborgs: Minds, Technologies, and the Future of Human Intelligence. Oxford University Press, Oxford (2003)

Corbin, J., Strauss, A.: Basics of Qualitative Research: Techniques and Procedures for Developing Grounded Theory, 3rd edn. Sage, Thousand Oaks (2008)

Deutsch, M.: Experimental philosophy and the theory of reference. Mind Lang. **24**(4), 445–466 (2009)

Dilthey, W.: The formation of the historical world in the human sciences. In: Makkreel, R.A., Rodi, F. (trans.). Princeton University Press, Princeton (2002, Original work published 1910)

Duhem, P.: Les Théories électrique de J. Clerk Maxwell: Etude Historique et Critique. A. Hermann & Cie, Paris (1902)

Eisner, E.: On the art and science of qualitative research in psychology. In: Camic, P., Rhodes, J., Yardly, L. (eds.) Qualitative Research in Psychology, pp. 17–29. American Psychological Association, Washington, DC (2003)

Garfinkel, H: Studies in ethnomethodology. Prentice-Hall, Englewood Cliffs, NJ (1967)

Giere, R.: Scientific Perspectivism. University of Chicago Press, Chicago (2006)

Greeno, J.: The Situativity of knowing, learning, and research. Am. Psychol. **5**(1), 5–26 (1998)

Heimann, P.M.: Maxwell and the models of consistent representation. Arch. Hist. Exact Sci. **6**, 171–213 (1970)

Hesse, M.: Models and Analogies in Science. Sheed and Ward, London (1963)

Hutchins, E.: How a cockpit remembers its speeds. Cogn. Sci. **19**, 265–288 (1995a)

Hutchins, E.: Cognition in the Wild. MIT Press, Cambridge (1995b)

James, W.: The Principles of Psychology, vol. 1-2. Henry Holt, New York (1890)

Kastenhofer, K.: Two sides of the same coin? The (techno)epistemic cultures of systems and synthetic biology. Stud. Hist. Philos. Sci. Part C: Stud. Hist. Philos. Biol. Biomed. Sci. **44**(2), 130–140 (2013)

Knobe, J., Shaun, N. (eds.): Experimental Philosophy, vol. 2. Oxford University Press, Oxford (2013)

Knuuttila, T., Loettgers, A.: Synthetic Modeling and the Functional Role of Noise. In: Epistemology of Modeling and Simulation: Building Research Bridges between the Philosophical and Modeling Communities, Pittsburgh, 1–3 April 2011

Kuhn, T.: The Structure of Scientific Revolutions: International Encyclopedia of Unified Science. University of Chicago Press, Chicago (1962)

Lave, J.: Cognition in practice: Mind, mathematics and culture in everyday life. Cambridge University Press, Cambridge, (1988)

Lincoln, Y., Guba, E.: Naturalistic Inquiry. Sage, Newbury Park (1985)

Longino, H.: The fate of knowledge. Princeton University Press, Princeton, NJ (2002)

Machery, E., Mallon, R., Nichols, S., Stich, S.P.: Semantics, cross-cultural style. Cognition **92**(3), 1–12 (2004)

Mead, G.H.: Mind, self, and society from the perspective of a social behaviorist. University of Chicago, Chicago (1934)

Nersessian, N.J.: Faraday to Einstein: Constructing Meaning in Scientific Theories. Kluwer Academic Publishers, Dordrecht (1984)

Nersessian, N.J.: How do scientists think? Capturing the dynamics of conceptual change in science. In: Giere, R. (ed.) Cognitive Models of Science, pp. 3–44. University of Minnesota Press, Minneapolis (1992)

Nersessian, N.J.: Opening the black box: cognitive science and the history of science. Osiris **10**, 194–211 (1995)
Nersessian, N.J.: The cognitive basis of model-based reasoning in science. In: Carruthers, P., Stich, S., Siegal, M. (eds.) The Cognitive Basis of Science, pp. 133–153. Cambridge University Press, Cambridge (2002)
Nersessian, N.J.: Interpreting scientific and engineering practices: integrating the cognitive, social, and cultural dimensions. In: Gorman, M.E., Tweney, R.D., Gooding, D.C., Kincannon, A. P. (eds.) Scientific and Technological Thinking, pp. 17–56. Lawrence Erlbaum, Hillsdale (2005)
Nersessian, N.J.: The cognitive-cultural systems of the research laboratory. Organizational Studies **27**(1), 125–145 (2006)
Nersessian, N.J.: Creating Scientific Concepts. MIT Press, Cambridge (2008)
Nersessian, N.J.: Engineering concepts: the interplay between concept formation and modeling practices in bioengineering sciences. Mind Cult. Act. **19**, 222–239 (2012)
Nersessian, N.J., Kurz-Milcke, E., Newstetter, W., Davies, J.: Research laboratories as evolving distributed cognitive systems. In: Alterman, D., Kirsch, D. (eds.) Proceedings of the cognitive science society, vol. 25, pp. 857–862. Lawrence Erlbaum Associates, Hillsdale (2003)
Newell, A.: Physical symbol systems. Cognitive science **4**(2), 135–183 (1980)
Osbeck, L., Nersessian, N.J.: The distribution of representation. J. Theory Soc. Behav. **36**(2), 141–160 (2006)
Osbeck, L.M., Nersessian, N.J.: Beyond motivation and metaphor: 'scientific passions' and anthropomorphism. In: EPSA11 Perspectives and Foundational Problems in Philosophy of Science, pp. 455–466. Springer, Heidelberg (2013)
Osbeck, L., Nersessian, N., Malone, K., Newstetter, W.: Science as Psychology. Sense-Making and Identity in Science Practice. Cambridge University Press, New York (2011)
Parfit, D.: Reasons and Persons. Oxford University Press, Oxford (1984)
Polanyi, M.: Personal Knowledge: Towards a Post-critical Philosophy. University of Chicago Press, Chicago (1973, Original work published 1958)
Rouse, J.: Engaging Science. How to Understand its Practices Philosophically. Cornell University Press, Ithaca (1996)
Rudwick, M.J.S.: The Great Devonian Controversy. University of Chicago Press, Chicago (1985)
Titchener, E.B.: The postulates of a structural psychology. Philos. Rev. **7**, 449–465 (1898)
Titchener, E.B.: Structural and functional psychology. Philos. Rev. **8**, 290–299 (1899)
Titchener, E.B.: The schema of introspection. Am. J. Psychol. **23**, 485–508 (1912)
Turner, J.C.: Toward a cognitive redefinition of the social group. In: Tajfel, H. (ed.) Social Identity and Intergroup Relations, pp. 93–118. Cambridge University Press, Cambridge (1982)
Vygotsky, L.: Mind and society: The development of higher mental processes. Cambridge, MA, Harvard University Press (1978)
Wertz, F., Charmaz, K., McMullen, L. Josselson, R., Anderson, R., McSpadden. E.: Five ways of doing qualitative analysis. Guilford, New York (2011)
Wundt, W.: Völkerpsychologie, vol. 1. Engelmann, Leipzig (1901)

Feeling with the Organism: A Blueprint for an Empirical Philosophy of Science

Erika Mansnerus and Susann Wagenknecht

Abstract Empirical insights have proven fruitful for the advancement of Philosophy of Science, but the integration of philosophical concepts and qualitative empirical data poses considerable methodological challenges. Debates in Integrated History and Philosophy of Science suggest that the advancement of philosophical knowledge can take place through the integration of empirical or historical research into philosophical studies, as Chang, Nersessian, Thagard and Schickore argue. Building upon their contributions, we will develop a blueprint for an *Empirical Philosophy of Science* that draws upon qualitative methods from the social sciences in order to advance our philosophical understanding of science in practice. We will regard the relationship between philosophical conceptualization and empirical data as an *iterative dialogue between theory and data,* which is guided by a particular '*feeling with*' the empirical phenomenon under study. On the basis of our own experience, we will explain how this dialogical interplay between conceptual discourse and empirical insight manifests itself when analysing the practices of infectious disease modelling and a team of planetary scientists. Thereby, we offer not only practical examples, but also a framework for further reflection on the methodology of an Empirical Philosophy of Science.

Keywords Philosophy of science in practice · Case studies · Qualitative research · Dialogue · Iteration · Social epistemology

---

E. Mansnerus
Department of Social Policy, London School of Economics and Political Science, LSE Health and Social Care, Houghton Street, London, UK

S. Wagenknecht (✉)
Centre for Science Studies, Aarhus University, Aarhus, Denmark
e-mail: su.wagen@gmail.com

1 Introduction

Empirical insights are increasingly valued in Philosophy of Science, and the growing interest in an understanding of science as practice, based on empirical case studies, can lead us to re-focus philosophical and epistemological questions (e.g. Ankeny et al. 2011). The focus on scientific practice as empirically accessible is supported by a 'naturalized philosophy' that recognises the need to accommodate philosophical analyses with empirical findings and methods (cf. Giere 1988; Kitcher 1992; Wylie 2002). In a similar vein, recent developments such as a 'socially relevant Philosophy of Science' seek to formulate an empirically-informed account of science and advance the understanding of science through interaction with scientists (e.g. Douglas 2010; Fehr and Plaisance 2010). Yet, little attention has been given to the question of how empirical research methods can serve philosophical analysis. By exploring how we actually collect, treat and reflect on first-hand empirical insights in relation to Philosophy of Science, we will suggest a blueprint for an *Empirical Philosophy of Science*.[1]

By way of an introduction, we revisit the long-standing debate on the relation between theory and empirical data that has unfolded at the interfache of History and Philosophy of Science. This debate discusses the possibility of an integrated History and Philosophy of Science (iHPS), a field that is envisioned as combining historical insight and philosophical theorizing by some—a program that is contested by others. We do not seek to contribute to the debate on iHPS ourselves. But we see informative parallels between iHPS and an Empirical Philosophy of Science that draws upon data from qualitative social-scientific methods, parallels also drawn by other authors, notably Henrik Thorén, in this volume. Returning to existing arguments for and against iHPS will help outline in which way an Empirical Philosophy of Science may, and may not, proceed.

In revisiting the debate on History and Philosophy of Science, Hasok Chang identifies its methodological dilemma: The historian-philosopher is faced with the choice between "making unwarranted generalizations from historical cases and doing entirely 'local' histories with no bearing on an overall understanding of the scientific process" (Chang 2012, p. 110). Still, Chang argues, we need not resort to the sceptical caution with which Thomas Kuhn, at later stages of his career, approached the project of an integrated History and Philosophy of Science. In

---

[1] *Empirical philosophy* as a concept is not well established in the discourse of philosophy of science. But in her study on the multiple perspectives and practices involved in the treatment of atherosclerosis as a disease, Mol (2002) argues for an 'empirical philosophy' that employs ethnographic methods and interests in order to address epistemological questions in knowledge practices. In a similar vein, we introduce qualitative empirical insights to the philosophical discourses, such as nature of scientific models.

dispute with Lakatos,[2] Kuhn emphatically defends the autonomy of history vis-à-vis Philosophy of Science. According to Kuhn, history of science is best written in the absence of philosophy, since historians should approach their empirical data without the presuppositions that a philosophical framework would impose: "The historian's problem is not simply that the facts do not speak for themselves but that [...] they speak exceedingly softly. Quiet is required if they are to be heard at all" (Kuhn 1980, p. 183). As a consequence, Kuhn is rather sceptical towards efforts to "amalgamate history and philosophy of science" (ibid.), but neither Chang nor we share his scepticism. In Chang's view, the dilemma appears to be severe only in the language of inductivism. With an inductivist perspective on an integrated History and Philosophy of Science, the historian-philosopher cannot avoid to either make unwarranted generalizations or to refrain from any generalization at all. Chang, however, proposes another perspective on an integrated History and Philosophy of Science:

> In attempting to transcend this dilemma, I believe that the first thing we need to do is to see if we can get beyond and inductive view of the history-philosophy relation, which takes history as particular and philosophy as general. Of course we cannot get away from inductive thinking entirely, but it is instructive to try seeing the history-philosophy relation as one between the *concrete* and the *abstract*, instead of one between the particular and the general (Chang 2012, p. 110; italics i.o.).

While the concrete pertains to the specific, idiosyncratic and particular, the abstract pertains to the conceptual that can be attached to different specific episodes or contexts. In any description of analytical depth, the abstract and the concrete will necessarily occur inextricably interwoven: "This necessity should not be resisted or avoided, but actively embraced as a great intellectual opportunity" (Chang 2012, p. 111). The abstract and the concrete give meaning to one another; and it is no undue generalization to carefully discuss in how far the abstract, as it has been articulated in one concrete context, applies to another context.

In our experience, the abstract is a rather heterogeneous conceptual component. Even in decidedly philosophical accounts, the abstracts will comprise *more* than philosophical concepts. Basic abstract ideas come with our language, with metaphors, with 'folk theories' about the social, with the popular (social) scientific knowledge that we possess and with our expertise in fields other than Philosophy of Science. The philosophical concepts of interest are but one element of the abstract. They are not irreplaceable, and we can choose one philosophical concept over another or develop an altogether new one. Likewise, Pitt (2001) has argued that historical episodes can be contextualized in more than one way. Writing history is necessarily selective in that the historian chooses one conceptual framework out of many. It is a mistake "to give the impression that there is only one appropriate

---

[2]Lakatos famously has argued that historical case studies on science without philosophical conceptualization are "blind" (Lakatos 1971: 91). He thus has favoured a history of science under the patronage of Philosophy of Science. This position has been harshly criticized not the least by Kuhn: "What Lakatos conceives as history is not history at all but philosophy fabricating examples" (Kuhn 1971: 143).

context that satisfies the explanatory-allowing role [which a framework is supposed to facilitate, our addition]" (Pitt 2001, p. 377). Contrary to our standpoint, however, Pitt argues that the selectivity in historical work constitutes an insurmountable problem for the empirically inspired philosopher of science:

> On the one hand, if the case is selected because it exemplifies the philosophical point being articulated, then it is not clear that the philosophical claims have been supported, because it could be argued that the historical data was manipulated to fit the point. On the other hand, if one starts with a case study, it is not clear where to go from there—for it is unreasonable to generalize from one case to even two or three (Pitt 2001, p. 373).

We are convinced, unlike Pitt, that empirical case studies can advance Philosophy of Science conceptually without manipulating empirical data to fit philosophical concepts. Pitt's notion of 'generalization' is misleading, as it suggests that generalization necessarily amounts to the formulation of universal laws. In our view, however, a case-based Empirical Philosophy of Science does not seek to contribute to universal laws of or rules for good science. Instead, as Burian suggests, philosophers "[…] must work in, and study, particular contexts and do our best to find valid, but limited generalizations" (Burian 2001, p. 399).

In reaching limited generalizations, empirically engaged philosophers can avoid unwarranted generality and undue data manipulation when they create a *dialogue between the abstract and the concrete* that is loyal to the phenomenon under study. The abstract need not be, and must not be, forced upon the concrete. Rather, the abstract should develop along with the concrete. The choice of particular philosophical concepts (as one component of the abstract) should enlighten the subject matter at hand. And while the abstract should shed light on the concrete, the concrete should force us to reconsider the abstract. It is this productive interplay, the dialoguing between the abstract and the concrete, which prevents us from fitting empirical insights unduly to philosophical concepts.

The critique of an inductivist perspective on case studies and empirical insights is not new in Philosophy of Science. Philosophers who work with case studies have felt the need to sketch out similar positions before: The use of empirical insights, Nersessian argues, should be by no means as understood as making "simple inductions" and thus being logically flawed and philosophically illegitimate (Nersessian 1991, p. 683). In fact, case-study based reasoning in Philosophy of Science is best understood as a "bootstrap procedure" in the course of which hypothesis are formulated on the basis of a theoretical background, then refined with and continuously tested against empirical insights from case studies (ibid.). This method will not "generate sweeping and comprehensive theories of science" (ibid, p. 684), but deliver theoretical accounts of science as a heterogeneous enterprise. Bootstrapping procedures, working back and forth between data and theory, are open-ended in principle. They find an end when a "reflective equilibrium" (Thagard 1988, p. 119) is reached. In other words, case-based analyses in Philosophy and History of Science (and, we argue, in an Empirical Philosophy of Science as well) unfurl in iterative loops of interpretation. They cannot comply with a simplistic inductivist view on the relationship between philosophical theory and empirical data, conceived of as 'evidence,' in which the latter seeks to 'confront' the former. This "confrontation model" is, as Schickore (2011) points out, highly problematic, since it construes an empirically engaged Philosophy of Science as analogous to a naïve picture of natural science.

Building upon this literature, we argue in this chapter that the empirically engaged philosopher should move back and forth between abstract and concrete, between concept formation and fieldwork. Crucial for the quality of the interplay between the conceptual and empirical insights is a close relationship to the empirical phenomenon under study, a relationship which we characterize with reference to Evelyn Fox Keller's description of Barbara McClintock's work as a 'feeling with the organism' (Fox Keller 1983, pp. xiii-xiv). This 'feeling with' is what guides us fruitfully through the creation of and careful reflection on first-hand empirical insights. Kuhn has been right in pointing out that the empirical is delicate and its insights are easily silenced by philosophical theory—but, we argue, he has been wrong in concluding that empirical work is thus best carried out in the absence of theory. Empirical work does not necessarily require Kuhnian 'quiet.' Instead, it requires a commitment to, a 'feeling with the concrete phenomenon at hand'.

We will unfold our concept of 'feeling with' further in the following Sect. 2. There, we will discuss the notions that in our experience have proven helpful to describe the modus operandi, which we would recommend for an *Empirical Philosophy of Science*, elaborating on our notion of 'dialoguing' between philosophical concept formation and empirical insights. To reflect on methodological aspects of our work further, Sects. 3 and 4 will discuss the modus operandi of empirical work for philosophy that Erika Mansnerus and Susann Wagenknecht respectively have developed for their different purposes. While one approach will be primarily driven by empirical data, the other one will be theory-driven:

By immersing herself into the process of modelling infectious diseases in an interdisciplinary research centre, Mansnerus learned about the key principle that guided modellers' work, i.e. how to 'let the data speak for themselves'. This meant that in order to build a good model, they tried to get closer and closer to the dynamics of what was happening in the data, in the infectious disease transmission dynamics they tried to model. 'Letting the data speak for themselves' required that the modellers were open to learn about the infectious dynamics that guided them to choose the model design in the most appropriate way. This meant that the modelling was not primarily guided by ideas what is technically possible, but by ideas of what is meaningful to understand the disease transmission. In a similar vein, through several interviews and ethnographic observations both in their seminars and work meetings, Mansnerus began to listen more and more what they had to say about the modelling process. Her pre-constructed ideas gave way and she let the 'data speak for themselves'. This took place in a form of a dialogue in interdisciplinary research seminars where the modellers talked about their work. In this manner, Mansnerus has developed a closer and more intimate relationship with her research object than traditional Philosophy of Science would have let her.

In contrast to Mansnerus, Wagenknecht approaches the empirical with stronger theoretical guidance. In doing so, she mediates between the established conceptual discourse in philosophy, i.e. 'theory,' and the empirically observable. In her approach, philosophical conceptualization is understood as open dialoguing, mediated by the researcher who gives voice both to her data and existing concepts. This dialogue reveals relations of resonance as well as relations of dissonance between abstract and concrete, existing concepts and qualitative emprical data. In the result of this dialogue, philosophical concepts are modified, further differentiated or rejected.

We will summarize our suggestions for a blueprint of an Empirical Philosophy of Science that uses qualitative research methods in Sect. 5.

2 Towards a Blueprint of an Empirical Philosophy of Science

While philosophy certainly tends to cherish the abstract, adjacent disciplines that commonly work with empirical methods often tend to focus strongly on the concrete. The primacy of theory in philosophy might lead philosophers to construct 'thin' cases or to resort to 'second hand' cases. The commitment to empirical scrutiny in Science and Technology Studies (STS), on the other hand, might lead to analytical and conceptual deficits. If the discussion of theoretical frameworks is neglected, only 'shallow' concepts can be derived from empirical work. Neither 'thin' cases nor 'shallow' concepts will do much to advance the study of scientific practice. Instead, thorough empirical work *and* analytically deep vocabularies are needed. Therefore, our recommendation for a blueprint of an *Empirical Philosophy of Science* is twofold: First, we suggest that empirical work is best accompanied by a close commitment to the phenomenon under study—a commitment which we will in the following describe in analogy to the 'feeling for the organism' which Barbara McClintock developed in her research in plant biology (Sect. 2.1). Second, we propose to reconcile first-hand empirical insights and the philosophical discourse, its concepts and hypotheses, in a dialogue between abstract and concrete (Sect. 2.2).

2.1 McClintock's 'Feeling for the Organism'—Philosophers' Feeling with the Phenomenon Under Study

In her biography of Barbara McClintock, Nobel Prize winning biologist, Evelyn Fox Keller highlights McClintock's ability to develop a particular 'feeling for the organism' that she studies:

> McClintock has pushed her special blend of observational and cognitive skills so far that few can follow her. She herself cannot quite say how she "knows" what she knows. She talks about the limits of verbally explicit reasoning; she stresses the importance of her "feeling for the organism" in terms that sound like those of mysticism. But like all good mystics, she insists on the utmost critical rigor, and, like all good scientists, her understanding emerges from a thorough absorption in, even identification with, her material (Fox Keller 1983, pp. xiii–xiv).

Absorption, a recognition for the limits of explicit reasoning and the willingness to closely intertwine the observational with the cognitive are the key characteristics of McClintock's research practice, as Fox Keller describes it. They are expressed through the metaphor of *'feeling for the organism'*. How does this feeling manifest itself?

> What enabled McClintock to see further and deeper into the mysteries of genetics than her colleagues? Her answer is simple. Over and over again, she tells us one must have the time to look, the patience to *'hear what the material has to say to you,'* the openness to *'let it come to you.'* Above all, one must have a *'feeling for the organism'*. One must understand *"how it grows, understand its parts, understand when something is going wrong with it*. [An organism] isn't just a piece of plastic, it's something that is constantly being affected by the environment, constantly showing attributes or disabilities in its growth. You have to be aware of all of that… *You need to know those plants* well enough so that if anything changes, [...] you [can] look at the plant and right away you know what this damage you see is from—something that scarped across it or something that bit it or something that wind did." You need to have a feeling for every individual plant (Fox Keller 1983, pp. 197–198, our italics).

When we read her own words (as highlighted), we notice that for McClintock, her 'feeling for the organism' is an active, nearly dialogical relationship with her corn (which was the key plant she studied). To 'hear what the material has to say to you' or 'let it come to you' are at odds with a preconceived attitude towards the research object. Yet the demand for cognitive understanding is there as a requirement to understand the growth of the plant, its parts, and the moments when something is wrong. The cognitive grounding arises and complements what she identifies as intuitive closeness or a 'mystic' experience in her ways to unfold the secrets of corn. McClintock's 'feeling for the organism' is the result of physical, but also highly emotional *and* intellectual proximity.

Drawing upon Evelyn Fox Keller's biography of Barbara McClintock, Knorr-Cetina (1997) takes her 'feeling for the organism' to illustrate a particular kind of solidarity, a feeling of unity which can arise between researcher and research object. This solidarity comes, as Knorr-Cetina describes, with a form of reciprocal reflexivity between researcher and the object of her study. From the researcher's perspective, the object under study is an incomplete object, a lacking object and the researcher's unfulfilled desire to know is 'looped through the object and back' (ibid, p. 16). A sequence of such loops, then, can be understood as the kind of reciprocal reflexivity that arises between researcher and research object. Solidarity and reciprocal reflexivity cannot be achieved without close acquaintance, i.e. without intimate knowledge about the research object (ibid., p. 21).[3]

---

[3]Knorr-Cetina's interest in object-centred sociality stems from a broader concern for forms of sociality in highly individualized, contemporary societies. It is her intention to explore notions of sociality with the help of models which she regards as "metaphors or tools to try out on the problem at hand" (Knorr-Cetina 1997, p. 20). Her description of the researcher-object relationship is intended to be metaphorical. However, other accounts of scientific practice such as Polanyi's stress that emotions such as the intimate care expressed in Fox Keller's 'feeling for the organism' do in fact operate and are epistemically effective.

We take McClintock's 'feeling for the organism' as a case for intuitive, emotionally laden creativity whose role in scientific practice Polanyi has unremittingly underlined. He conceives of 'knowing' "[...] as an active comprehension of the things known," which comes with a personal commitment and is driven by "intellectual passions" (Polanyi 1962, p. vii). These passions guide the researcher's attention in that they have a selective and heuristic function (Polanyi 1962, p. 134, 142). In a similar vein, Ronald de Sousa has extensively argued that emotions underlie our rational processes. On his view, emotions direct our cognitive capacities in that they determine "patterns of salience among objects of attention, lines of inquiry, and inferential strategies" (de Sousa 1987, p. 196).

In their study of a biomedical engineering laboratory Osbeck et al. (2011) observe that scientists use an emotive language of care and responsibility when talking about their research objects. When interviewing practising scientists about their research objects, they have repeatedly encountered expressions of anthropomorphism among which the theme of 'making your cells happy' figures prominently. Attributing emotions of happiness to research objects helps the researcher to create a productive relation to her research object, and Osbeck et al. observe a "dynamic interplay between the attributed happiness of the cells and the researcher's cognitive goals" (Osbeck et al. 2011, p. 117). Interactive intimacy is a precondition for such a dynamic interplay. The theme of "making your cells happy" has thus a social and a normative component. It implies that the researcher is intensely interacting with and is responsible for the happiness of the cells she is cultivating (Osbeck et al. 2011, p. 114).

In analogy to the descriptions of scientific inquiry that Fox Keller, Osbeck et al. and others have provided, we will speak of a 'feeling *with*' the phenomenon under study as we outline our blueprint for an Empirical Philosophy of Science. In our perspective, a 'feeling with' the phenomenon is an attitude of care and commitment to the object (or, the subjects) under study. It is an attitude, we argue, that the empirical philosopher should adopt in her attempts to create a fruitful interaction between empirical insights and philosophical concept formation. A 'feeling with' the organism requires intellectual, emotional and physical proximity to the object under study. Hereby, physical presence should not be underestimated.[4] We argue that through an on-going, empathic involvement with the object of research at the research site, the researcher is able to develop a deeper understanding of how practicing scientists think and work, how they form collaborations and how they can produce good scientific knowledge.

---

[4]Collins (1991) sees a difference between historical and sociological studies of science in terms of *distance*. Historians are capable of distancing themselves from their objects of study due to the time-scale of their studies, whereas sociologists are able to provide in-depth insights based on their closeness to the object of study. So, in our empirical work, we make use of the potential offered by social scientific methods, which promote a reflected-on proximity between researcher and the phenomenon under study.

2.2 Dialoguing as Modus Operandi

The 'feeling with' a concrete research phenomenon at hand comes to effect during the research process in what we understand as a 'dialogue between the abstract and the concrete'. Based on our own research experience, we describe the modus operandi of an Empirical Philosophy of Science as 'dialoguing'. In our understanding, dialoguing *can take a number of different forms* ranging from an inter-personal dialogue between philosopher and scientists in the field to an inner dialogue at the philosopher's desk. Yet, in all these ways, dialogue is iterative by nature and describes a process of moving back and forth between two different perspectives or accounts.

Iteration, repeated moves of back and forth, are central to any dialoguing. In dialogues between the abstract and the concrete, empirical data and possible conceptualizations are constantly revisited in order to explore how, if and why (not) they can bear on one another. Epistemic iteration as a feature of the course in which experimental science can proceed describes the non-foundationalist and non-inductivist development of empirical knowledge, as illustrated in Chang's account of temperature (Chang 2004). Iterative dialoguing for an empirical philosophy seeks to create 'feedback loops' between abstract and concrete to develop concepts, which are deeply rooted both in theory and in data. Such feedback can take place in the empirical philosopher's reasoning. It can, however, also take the form of feedback from the people observed (Hacking 1995).

Another crucial element of dialoguing is a repeated change of perspectives. In order to better account for science as practiced by human agents, Chang has argued that scientific practice is well understood as dialoguing activity from a first-person/second-person perspective (Chang 2011, p. 212). Knowledge creation as epistemic practice is, Chang points out, essentially a dialogue between 'me' and 'you'. Chang refers in his reflections to Buber's philosophy of dialogue (cf. Buber 1970), which highlights "I-Thou" vs. "I-It" distinction. Whereas an "I-Thou" relationship features a "direct, mutual, present and open" dialogue,[5] the "I-It" relationship rests on a monologous "[…] relation, in which one relates to the other only indirectly and non-mutually, knowing and using the other" (Friedman 1955, p. 26). In contrast to a monologue, a genuine dialogue thus addresses the other as second person and is actively engaged with her (Friedman 1955, p. 37).

We use the ideal of an iterative dialogue between 'I' and 'You' as a methodological directive, as a heuristic tool and as resource of methodological reflection when mobilizing qualitative data for Philosophy of Science. When we characterize *Empirical Philosophy of Science* as a dialogue, we seek to emphasize

---

[5]Buber takes a rather radical standpoint on his "I-Thou" relationship: "The relation to the You is unmediated. Nothing conceptual intervenes between I and You, no prior knowledge and no imagination […]" (Buber 1970, p. 62). Clearly, we do not follow Buber in this regard, as we regard the abstract, i.e. the conceptual broadly conceived, as integral part of those experiences which matter to empirical fieldwork.

methodological reflection on three aspects: First, dialogues are *open-ended*. Interpretive empirical research unfolds over time and the course of its iterations is difficult to foresee. Second, dialogues are *situated* spatially, temporally, socially, culturally, institutionally and discursively. This, however, does not imply that the results of an Empirical Philosophy of Science would be arbitrary. On the contrary, they are not arbitrary, precisely because they are rooted in a situated research process. Third, dialogues live of *differences*. They consist in a back and forth between at least two differently angled viewpoints. Dialogues seldom result in the fusion of viewpoints; rather, they feature dissonances, clashes, frictions, partial consensus. Ideally, dialogues acknowledge difference. Neither need the philosopher accept uncritically the scientist's viewpoint, nor need she impose her view upon the scientist. Neither need we silence the existing philosophical discourse, nor need we brush over the wealth, depth and heterogeneity of qualitative data. Neither need the abstract be sweepingly invalidated by the concrete, nor need the concrete be overruled by the abstract.

Let us conclude this section by pointing out that the dialogues of qualitative method within and for Philosophy of Science are to a substantial extent internalized. They are often 'inner dialogues,' initiated and driven by the researcher who voices both the empirical and the theoretical. Such inner dialoguing involves sophisticated moves of turn-taking and is flexible in its accentuation. The shifting accent in dialoguing—either on the abstract and more specifically of philosophical concepts, or on the concrete—can lead to different acts of conceptualization. Whereas Mansnerus rooted her dialoguing decidedly in the concrete, moved gradually to the abstract and took philosophical concepts into account only later, Wagenknecht followed a different trajectory. Her work can be seen as involving the abstract, especially concepts from Philosophy of Science, right from the start.

3 Let the Data Speak for Themselves

Erika Mansnerus

The leading idea throughout my research was to let the empirical material show the way—to let the data speak for themselves. I applied the so called Laboratory Studies tradition within Science and Technology Studies. These approaches identified themselves as anthropologies of knowledge or anthropology of science. If we look at these early studies in sociology of science, we find appreciation of a *sensitivity* towards the research process. The process acknowledges a preliminary presentation of accumulated empirical material. This means studying the research procedure undertaken by research scientists, and it often relies on documentation of the research process. When anthropologists apply techniques of participatory observation, they collect data and describe the scientific activity from a more personal point of view. What this method aims at is a more comprehensive understanding of both the technical and social aspects of scientific activity. The technical (instruments, laboratory procedures, experimental practices) is equally important to the social and interactional side of science (e.g. Knorr-Cetina 1981; Latour and Woolgar 1986). This schematic idea of how ethnographical study is conducted in laboratory settings guided my research and it was enriched by "socio-logic of research," a process of studying both the social and epistemic development of science (Callon 1980). So I embraced the "socio-logic", the motivation to understand, not only the social, organisational and interdisciplinary dynamics of modelling research, but the epistemic, the models themselves.

For me, the purpose of ethnographic research was to engage in a multi-voiced method that allowed me to incorporate both interview data and participatory observation. Ethnography has a narrative element that allows combining experiences and textual description in order to reach "thick description" of the studied phenomenon (e.g. Geertz 1973/2001). As such, ethnographic knowledge remains subjective, contextual and partial, bound to researcher's perspectives and standpoints, yet my aspiration was to engage with philosophical analysis as well.

Familiar with laboratories, I decided for a change to observe a field expedition. I also decided, 'being something of a philosopher,' to use my report on the expedition as a chance to study empirically the epistemological question of scientific reference (Latour 1999, p. 26).

Bruno Latour's self-expression of "being something of a philosopher" resonates with the uneasiness of combining qualitative empirical research with philosophical inquiry. Is that even possible, one might ask? This section examines how empirical material is brought into dialogue with philosophical questions and by doing so, I will reflect on my process of conducting 'anthropology of infectious disease modelling,' which arises from my understanding of ethnography as a method to conduct qualitative research. Ethnographic understanding is produced in dialogue between the researcher and those who have been studied. This *dialogue* led me to be inspired by the idea of 'let the data speak for themselves' that was an initial description of the modelling practices I learned about. How did it emerge through my empirical research?

My case study was to reconstruct a life span of an interdisciplinary infectious disease modelling project that took place as a collaboration between the University of Helsinki, Technical University of Helsinki[6] and The National Institute of Public Health.[7] I participated in their regular research seminars, interviewed the key actors, but also studied their models over a period of two years. In particular, I followed a series of work meetings in which two simulation models on *Haemophilus influenzae* type b (Hib) transmission and vaccination effects were built. Through the long-term contact with the modellers, I learned how their way into the epidemiological research happened. They described the process as 'let the data speak for themselves.' This was a guiding principle through which they learned what the epidemiological data actually represented and how that knowledge could guide them in model parameterisation. They tried to understand the epidemiological

---

[6]Currently the Aalto University of Technology.
[7]Currently the National Institute for Health and Welfare.

processes as realistically as possible. I was inspired by this—how could I 'let the data speak for themselves' in my study? My close connection with the modellers and epidemiologists gave me new perspectives, taught me how to read the models and what kind of debates model-building contained. I wanted to create a dialogue with the modellers and invited them to give a talk in a seminar that was aimed at science studies scholars. A fruitful concept to describe modelling as tailoring arose from that dialogue. Later on, this dialogue has led to research collaboration beyond my PhD project and the modellers have usefully adopted an open way to communicate their key research findings.

3.1 Tailoring a Model

The metaphor of tailoring was brought alive by a senior scientist, who clarified the practice itself in a research seminar, at the University of Helsinki:

> I think something that would be closer to tailoring that is intending or trying to make a suit for a client. Because, I mean here we have the materials, here we have some tools, and it's not just the tools that we should be describing. Here we have a client or a goal and we do something towards the goal (Senior researcher 5.10.2001).

Tailoring, as he explained, is based on a good relationship between tailor and customer; it requires on-going communication of the customer's wishes and needs, which are then expanded upon in 'fitting sessions'; the final result is a specifically designed suit to fit a particular client. A model might be similarly tailored to answer policy calls from public health officials. This metaphor he used did not exist in the philosophical or sociological discourses on modelling. It emerged in a joint seminar in which both the modellers and myself gave presentations. As a reflection of their work, the modellers identified and characterised that as tailoring. I let the data speak for themselves by organising this seminar and inviting their contributions. In a similar way, they let the data speak for themselves in the several work meetings I observed. The interdisciplinary research collaboration brought the modellers in a regular working contact with the epidemiologists. In those discussions, what was possible to express mathematically was measured against its epidemiological rationale (cf. Mattila 2006). I realised that through the openness of understanding models and modelling practices, I had to bring the concept to the philosophical and sociological realms. I did that by defining tailoring as building models with the intention for use. In this sense, the notion arose from the concrete description of modelling work and gave an abstraction of a particular modelling practice. Theoretically, the notion relied on the understanding that models are seen as mediators, instruments in scientific work and as such they function as autonomous objects (Morgan and Morrison 1999).

How to work towards philosophical questions when one is immersed in ethnographic research? In order to reflect on the empirical research process, I would like to look at two concepts that arose from the dialogue with the modellers, which took place in interviews and in the joint seminar. While conducting my study, the philosophical discussion on the roles and functions of scientific models became vivid (cf. Sismondo 1999; Morgan and Morrison 1999). Here, key arguments reassessed models as scientific instruments rather than theories, acknowledged that models are capable of functioning in as measuring devices, substitutes for experiments or mediating scientific work. Models were increasingly seen as 'social constructions.' Their multiplex and unfolding nature as inhabitants of experimental worlds informed philosophical and sociological work on models. Yet, these accounts left their applicability intact. Tailoring as a concept for and description of given modelling practices was a way to work forward. Tailoring highlighted the key aspect of epidemiological modelling: to build models for applicable purpose (i.e. for public health research and policy). By introducing this concept into the philosophically driven debate, I enriched the conceptual framework by empirical understanding.

Another way of looking at the process in which the data spoke for themselves is to relate the process with the notion of iteration.[8] In this case, we can identify two levels of iteration. Iteration as a technical term that happens within the modelling practice, and iteration as an epistemic process, accumulating knowledge. Kari Auranen articulates iteration in the technical sense in his approach to modelling as an iterative exercise. In his words "a balance was sought between a realist enough description of the phenomenon under study and the amount of information and the sampling scheme of the data" during a modelling process (Auranen 1999, p. 16). This "seeking for a balance" shows how iteration happens between the data and data sampling and efforts to give a description of the phenomenon under study.

As discussed earlier, iteration has gained interest among philosophers of science through Chang's (2004) notion of epistemic iteration, which represents the second level of iteration. The iterative process itself, he argues is a process "in which we throw very imperfect ingredients together and manufacture something just a bit less imperfect." A difference between epistemic and methodological iteration can be identified (Elliott 2012). Methodological iteration can, according to Elliott, assist in developing a starting point to enrich further inquiry; to isolate and clarify problems with current knowledge claims; and to alter and enrich former knowledge claims when moving between the modes of research.

I argue that both epistemic and methodological iteration were present in my own process of making sense of epidemiological modelling and the nature of interdisciplinary environment in which modelling took place. Methodologically, I would agree that as I gained experience in my research, I was able to produce "improved epistemic outcomes" (cf. O'Malley 2011). Epistemically, my understanding of the successive stages of knowledge of models and modelling practices were tinkered through the ethnographic observations, various forms of communication with the

---

[8]Iteration itself means "repetition of an action or process"; repeated performance, if we follow a dictionary definition (OED).

modellers and by studying the models themselves (through the publications and manuscript drafts made available for me during the meetings).

The key challenge for my study was how to relate the subjective ethnographic understanding to Philosophy of Science. My research was in the grey zone of philosophy and sociology of science and I drew together resources from laboratory studies, naturalist Philosophy of Science and the so called practice turn in science studies, as discussed in the previous section. Yet it was not straightforward to say that this piece of research was philosophy. The logic of ethnography and the idea of dialogue felt incompatible with analytical and normative goals of philosophical argumentation. I didn't have Latour's authority to be playful and be something of a philosopher. This dilemma shadowed the process of finishing my PhD, but it was more than an administrative or organisational dilemma. In the end, the question remained: Can empirical research be relevant to Philosophy of Science?

3.2 From the Concrete to the Abstract: Artificial Nature

Having immersed myself into the details of models, I became familiar with the fast growing debates about their nature and standing in Philosophy of Science, I was looking for a conceptualisation that would capture the concrete practices involved in modelling as well as their philosophical significance. Knowing how a model is being built, through steps that usually begin by asking a question, building a model as an answer to it, parameterising a model and manipulating the model (e.g. in Morgan 2001), I began to think of models as *Artificial Nature*. This concept had two roots. The modellers I studied referred to their use of models when they were building them as a 'playground,'[9] which meant that they manipulated the possible, artificial worlds inside the model. These worlds captured different vaccination scenarios (e.g. different age cohorts being vaccinated, different vaccines given). In the playground, the modellers gained sense of how the model works and whether it produces 'mistakes' which could be due to calibration of the model or a bug in the programming.

With the feeling of manipulability of models, I looked into the philosophical traditions that emphasise questions as manipulative tools in experimental practices. The predecessors of this date back to Francis Bacon, and the early days of the experimental method, as the following quotation shows:

> To observe is to detect the actions of nature; but we shall not advance far in this path, unless we have a notion of its character. To make experiments is to lay questions before nature; but he who alone can do that beneficially knows what he should ask (Christian Oersted in 1852, quoted from Sintonen 2004).

---

[9]A mathematical modeller and an epidemiologist said that "we are able to play with models," which depicted their close collaboration and mutual understanding (cf. Mattila 2006).

"Laying questions before the nature" or "twist the lion's tail to manipulate our world in order to learn its secrets" as Hacking (1983, p. 149) describes the Baconian revolution, led me to see models and simulations as *Artificial Nature* that we can manipulate by questions. This allowed me to affiliate my approach with the *Interrogative model of inquiry*, which developed as a heuristic to understand knowledge-seeking processes either in logical or explanatory terms (cf. Hakkarainen and Sintonen 2002). As *Artificial Nature* models offered an indirect way to manipulate, experiment and question the choices and scenarios built into them.

To conclude, I suggest that the metaphor of 'letting the data speak for themselves' describes the iterative process of combining empirical research with philosophical endeavours. As my data spoke to me through the ethnographic observations in seminars and work meetings, through interviews and through personal feedback, I learned to listen to it. But the data also spoke through the publications. Over time I learned to read the models, identify the assumptions made and see how the key results were discussed. The models as well as the modellers spoke, if I listened. This iterative approach challenges pre-existing assumptions and hypotheses that might have been indirectly guiding my research process. As a dynamic process, research that acknowledges and benefits from iteration opens to the possibilities that are not yet known. Instead of testing hypothesis, I let the novel observations unfold, and saw the value of describing modelling as *tailoring*, for example. The challenge for empirically driven research naturally comes from the philosophical understanding of science. In order to relate back to the philosophical debates, the empirically informed accounts need to be revisited. Fortunately, within the philosophy of modelling the path was cleared with the rich discourses of functions of models in scientific work. The importance of the empirical was even recognised in the idea of a 'motle' epistemology of modelling (Winsberg 1999, p. 275). All this allowed me to develop my practice-based views on modelling and bring them into dialogue with the philosophical accounts of scientific modelling (cf. Morgan and Morrison 1999).

4 Theorizing as Dialoguing

Susann Wagenknecht

In my work on the collaborative creation of scientific knowledge, I have started with a theoretical background and a theoretically inspired interest rooted at the intersection of Social Epistemology and Philosophy of Science. My empirical study of the division of labor in research groups has led me to distinguish different forms of epistemic dependence (Wagenknecht 2014). Moreover, I have formulated a contextualized account of epistemic trust in scientific practice that modifies Hardwig's (1985, 1991) perspective on trust among scientists on the basis of empirical insights (Wagenknecht 2015). In doing so, I have been able to explicate the interactive epistemic practices that underlie the creation of 'collective' knowledge. So far, the question whether and to what extent collaboratively created knowledge is to be described as genuinely 'collective' knowledge has almost exclusively been dealt with on purely conceptual grounds, by discussing hypothetical, supposedly 'typical' or generic examples and/or by drawing on common sense knowledge (Gilbert 2000; Wray 2002; Fagan 2011). The discussion of 'collective' knowledge has, however, not been systematically informed by the perspective of practicing scientists, a notable exception being the work of Staley (2007) and Rehg and Staley (2008) who rest their reflection on an interview study.

My approach to empirical methods for and within Philosophy of Science exemplifies a dialogue between abstract and concrete, philosophical vocabulary and empirical insights which originates from philosophical focus points—and which, eventually, seeks to develop philosophy's conceptual tools. Therefore, existing conceptual approaches strongly shape my research interest. My exploration of theory and observable social reality took place simultaneously in an on-going back and forth between literature review and theoretical reflection on the one hand and empirical work on the other. In doing so, my analytic conceptualizations and theoretical arguments co-evolved with my empirical inquiry in group research.

With an accentuated conceptual focus right from the start, there is clearly the danger of succumbing to an all-too-human confirmation bias and create a premature fit between established theoretical discourse and newly gained empirical insight. For this reason it is important not to corroborate any working hypothesis prematurely, but to explore both instances of *resonance and dissonance* between data and theory. This, however, requires making empirical data as 'strong' as long-established theoretical approaches are. Popular theoretical approaches have a whole range of adherents and defendants. Empirical data on the contrary are, in the first place, voiced only by the researcher who created them. Compared to philosophical concepts with a discursive tradition of five, ten or even more years, empirical data speak with a 'feeble voice.' I have experienced this imbalance as a constant challenge to nurture the necessary empirical sensitivity against the pressure of time and the cool elegance of philosophy's theories. Therefore, dedicated care for and thorough commitment to empirical data are important. Empirical data need an intellectual environment to unfold and thrive. They, too, need to be 'kept happy.' In the following Sects. 4.1 and 4.2, I will show in which ways qualitative empirical data can be investigated in their own right while their investigation is geared towards a previously established conceptual focus at the same time.

4.1 Data Generation with a Theoretical Starting Point

I have approached fieldwork, interviewing and the empirical data collected throughout these activities with a set of theoretical questions tied to philosophical concepts, such as: *How is epistemic dependence dealt with by practicing researchers? What forms and configurations of epistemic dependence are relevant in scientific practice? What is the nature of epistemic trust in practice? How is trust managed in research collaborations?* These questions were formulated—albeit preliminarily—prior to the generation of empirical data.

In order to answer these questions, I have comparatively studied two Danish research teams in different fields. In each group, my observations stretched roughly over the course of a year. At the beginning of my field work, I observed group meetings. Especially in the start, these observations are in parts the observations of a philosopher and in parts those of a plain visitor. I was an outsider. This, however, changed partly around the time when I started 'shadowing' them for one or several working days in a row. 'Shadowing' has been developed in modern ethnography (Czarniawska 2007) as a technique for following single persons through their daily life. It enabled me to focus on single group members. Although the content of their research is not what I have focused on, I have tried to learn about their scientific work as much as possible. I have read a number of their papers and asked them to explain basic experimental procedures. While following them through their laboratories, meetings and lunch breaks, we had conversations about their work. It was in these conversations that I started to develop a 'feeling' for them—especially when listening to the emotional undertone to their words. Nevertheless, there remained a professional slack between me and them. This slack helped to establish myself as a trustworthy outsider and opened a space for reflection which I otherwise would not have had.

Later on, when I had familiarized myself with the groups, I interviewed selected group members.[10] The shift from observing to interviewing was significant, since it implied a shift from a rather passive role in which I could keep my observations to myself to a role in which I structured my exchange with the scientists more openly. In the light of this experience, I suggest to conceive of interviewing as a co-construction between interviewer and interviewee (King and Horrocks 2010, p. 134), and I will elaborate on the co-constructive character of data collection through interviewing in the following.

For both interviewed scientist and interviewing philosopher, interviewing means a very intense confrontation with their own work in a compressed way: for the scientist, because she is asked to explicate a precise account of her work as a whole; for the interviewer, because every interview means testing out the fruitfulness of her research question. Therefore, interviewing mobilizes intellectual resources of both interviewee and interviewer. This gains relevance especially in highly asymmetrical interviewing situations such as the ones I have encountered. A philosopher-interviewer is not a fellow scientist and cannot pretend to be one. A scientist, in turn, is no philosopher. Both have their own knowledge about the phenomenon in question, be it theoretical foreknowledge or practical experience, and both kinds of knowledge are necessary to establish an interview relationship between them. Interviewing can be understood as mediation between these two different stocks of knowledge.

---

[10]On the integration of observation with interviewing for qualitative empirical inquiry see e.g. Coffey and Atkinson (1996).

The challenge, however, for any interviewing with philosophical intentions lies before the actual interviewing. Preparing good questions is key. Useful questions have to be both meaningful to the interviewee and meaningful with regard to the philosophical issues that the interview should contribute to. The gap between scientists' life worlds and Philosophy of Science is a gap not easily crossed, and it is the philosopher-interviewer who has to find access to the life worlds of her interviewees by rephrasing her philosophically motivated questions in a language they find understandable. In my experience, asking practicing scientists to take up philosophical theorizing does not yield results of the quality desired.[11]

For the ten interviews that I made in two research groups, I employed a semi-structured question format (cf. Fontana and Frey 2000, p. 653). I had a prepared interview guide with about twenty questions with me when I went to interviews. Sometimes I asked the questions I had prepared literally; other times I reformulated them ad hoc so that they would not interrupt the conversational flow. Often I changed their order. Since I had observed their work before, we could relate to specific persons, incidents or articles in our interview as to illuminate more general points. I regarded my interviewing as follow-up of my observations and tried to establish continuity between fieldwork and interviewing.

Certainly, I have interviewed 'experts' and my approach bears strong resemblance to approaches that Bogner and Menz (2009, pp. 47-48) characterize as "theory-generating expert interviewing." Nevertheless, I did not perceive of my interviewing as 'expert interviewing' as described by e.g. Zuckerman (1972). In fact, the relationship between interviewees and me was rather collegial and the interview atmosphere usually promoted a rather informal conversation. Often, interviewees would approach me as an 'expert for philosophy.' I, in turn, regarded *every* interviewee as 'expert' for her individual daily professional practice—notwithstanding age or reputation.

The concept of co-construction enables me to make sense of the fundamental asymmetries involved in interviewing. Interviewers as well as interviewees contribute to the interview, but they do so in very different ways. While the philosopher-interviewer should primarily be listening during the actual interview, the scientist-interviewee remains, except for occasional feedback, largely silent during the analysis process. Both interview partners are observers, i.e., they both observe and interpret what is happening during an interview, but the interpretation that is pivotal for a philosophical study is the one made by the philosopher. She 'keeps the rule'; the dialogue between abstract and concrete remains 'hers' and it remains, for the time being, within the realm of philosophy.[12]

---

[11]I have not gone so far as to employ openly 'collaborative' interviewing as described by Ellis and Berger (2003). I have restrained myself to asking questions, elaborating on these questions and offering reformulations. In single instances have I explained in simple terms how 'some philosophers would think about' the issue in questions. I have not, however, confronted interviewees with an elaborate description of my tentative, theoretically informed perspective.

[12]In contrast to e.g. Hasu and Miettinen (2006), my dialogical approach carries no 'interventionist' motivation.

4.2 Analysing Data in Dialogue with Theory

My analysis certainly unfolded vis-à-vis theoretical concepts and accounts. However, a theoretical focus should not force a particular interpretation upon empirical data and degrade empirical data to illustrations of existing theoretical approaches. To refrain from doing so, it is essential to consider empirical data in their own right. Fortunately, qualitative data, particularly interview data, have a number of features which can obstruct their premature subsumption under philosophical conceptions: Their abundance, depth, fine-grained structure and not least their linguistic qualities help to avoid tuning empirical data all-too-easily to any philosophical argument. Interviewees typically use a language that is not continuous with philosophy's technical terminology, and I suggest to regard this discrepancy as an encouragement to explore the deeper, multifarious ways in which empirical data and philosophical discourse can be related by the philosopher-investigator.

My commitment to philosophical questions has led me to direct my analysis of empirical data to a set of themes such as, e.g., relations of dependence, epistemic trust and the 'collective' nature of collaboratively created scientific knowledge.[13] In the beginning of my analysis, these themes were outlined loosely. Recasting the concepts around which the philosophical debate revolves in terms of empirically observable phenomena, I was able to assume a more distanced perspective upon the analytic subtleties of philosophical discourse. I did not approach my data with a distinct preference for one conceptual definition or philosophical argument over another. Rather, determinate analytic reflections emerged from the later phases of my data analysis. Therefore, I conceive of my approach to data analysis and interpretation neither as bottom-up nor as top-down approach, i.e., I have approached my data neither conceptually naïve nor with a rigid conceptual grid.

I have chosen to 'code' large parts of my data and combine coding techniques with phases in which I immersed myself in the data collected in a comparably unstructured manner. Being a term borrowed from grounded theory, to 'code' means to index text passages with labels for analytic purposes. The interpreter develops descriptive or/and analytic categories—so-called codes—and applies these codes to text passages while working her way through the text. Codes are "[…] conceptual labels placed on discrete happenings, events, and other instances of phenomena" (Strauss and Corbin 1990, p. 61; see also Alexa and Zuell 2000, p. 306). The investigator moves from indexing text passages and thereby familiarizing herself with them to conceptually reconstructing her data with regard to her own analytical focus. It is a process of familiarization and emancipation. By breaking up a body of text into manageable segments, describing those segments,

---

[13]For thematic analysis, in which a 'theme' "[…] captures something important about the data in relation to the research question, and represents some level of patterned response or meaning within the data set," see Braun and Clarke (2006, p. 82), but also Attride-Stirling (2001) and Boyatzis (1998).

comparing and relating them, the investigator analytically reworks her textual data such as to find answers to the questions that she approached her data with. My coding style can be described as moderately theory-directed "editing" of interview transcripts (Crabtree and Miller 1992, p. 94).

I did not formulate a template of codes prior to analysing. When I had transcribed my data gained from observing or interviewing, I started with applying ad hoc codes to semantic units in the text which seemed relevant to my research interest. I have used both rather descriptive codes which did not appear to be theoretically relevant at first glance and more interpretive, theory-inspired codes to label and organize text passages (cf. King and Horrocks 2010, p. 153; Strauss 1987, p. 33f.). An example for a strongly theory-guided code would be 'testimony.' Yet, labelling different text passages with this code has not helped me substantially in organizing my data material, because acts of testimony are simply all too ubiquitous in collaborative scientific practice. This insight led me to consider different forms of testimony. An example for a rather descriptive code which was, at least in the beginning, rather unconnected to my theoretical framework, would be 'frustration.' While shadowing a Ph.D. student, 'frustration' was a recurrent theme in his conversations with me. Clearly, his frustration was related to the high pressure to succeed, his anxiety to fail and his understanding that he, after all, had been given a high-risk project as dissertation topic. His frustration, however, was also related to his work conditions. He perceived his research group as "little interactive." This perception stood in sharp contrast to my observation and let me to reflect on my biases as an observer. Having a background in philosophy, the hustle and bustle of a biology laboratory must appear very "interactive" to me. I took his frustration as an occasion to study the interplay of delegation, help and individual responsibility in more detail.

At some points in the process, I decided to start developing templates regarding a specific issue I was investigating. This has helped me to systematize existing codes and envision new, complementing codes to match the existing ones. Yet, it has proven useful to combine highly structured phases of data analysis with repeated phases of unstructured immersion that have ensured that I would not 'lose touch' with the original data. So, in between coding cycles I have gone back to data in their raw, unprocessed form. After I had skimmed scribbled field notes again and re-listened intensively to original audio files, I wrote overall, encompassing case descriptions and composed 'holistic' characterizations of single interviews. This has helped me to check the validity of my interpreting attempts up to that point in time: *Did I distort what interviewees wanted to confer? Did I approach my data with all-too heavy theoretic machinery? Did I stay close enough to interviewees' reasoning?*

I experienced the process of my empirical work as a process with changing, sometimes ambivalent commitments. During my interviews with practicing scientists I felt as if I acted as the representative of theory, whereas I acted as the representative of the empirical at other times. When I was working on a rather abstract, conceptual level, I felt more committed to the empirically observable than to existing philosophical accounts. I thus regard my work as a constant dialogue between abstract and concrete in which I continuously swap roles. This dialogue is an open investigation of relations, resemblance, and resonance between the empirically accessible and the pre-existing theoretical sphere. It is a search for empirical insights and conceptual refinement at the same time. It may be a search for assumptions in philosophical accounts that can be shown to be incorrect for the empirical case at hand. Empirically-based conceptualizing in philosophy is, however, not a mere falsification tool. It is a constructive inquiry in which abstract philosophical arguments and empirical investigation benefit from one another, but are not merged into a homogeneous maze.

5 Feeling with the Organisms: A Blueprint for an Empirical Philosophy of Science

Our contribution is written as an invitation to make use of different qualitative methods to bear on concept formation in Philosophy of Science—and to reflect upon the methodology of a Philosophy of Science in Practice that draws on empirical work. To integrate ethnographic studies on the one hand and philosophical analyses of scientific practice on the other is a tall order. In our view, however, the dialoguing approach which we have elaborated in our contribution can provide a way forward.

To combine abstract philosophical concepts with empirical data is not a trivial endeavour. Often, rich empirical material does not correspond to philosophy's terminology and its subtleties. Faced with the gap between empirical data and the analytic vocabulary which philosophy offers, the empirical philosopher is challenged to establish a connection between these two. Thereby, she may run the risk of fitting empirical insights and philosophical framework unduly to each other. A 'feeling with' the phenomenon under study is our response to challenges such as this. We show how a 'feeling with' the phenomenon under study can inform an *Empirical Philosophy of Science* which seeks to ground philosophical conceptualization in first-hand empirical insight gained through qualitative case studies. A 'feeling with' designates the personal acquaintance and the attitude of commitment to the empirically studied, which arises through phases of physical, emotional and intellectual proximity throughout the process of empirical study. This process, we argue, unfolds ideally as an iterative dialogue between 'I' and 'you,' i.e. as a collaborative conversation between the philosopher who observes and interviews and the practicing scientists.

The dialogue between 'I' and 'you' concerns our behaviour in the field of practice which we seek to study and thus can be seen as the 'outer' logic of our empirical approach. The 'inner' logic of our approach, i.e. our philosophical reasoning, is aptly described as a dialogue between abstract and concrete. A 'feeling with,' we argue, guides us fruitfully in the interplay between abstract and concrete in which philosophical concepts and empirical data are brought to bear on one another. Theoretical work without a thorough empirical basis is maybe not empty, but risks developing into directions that are neither relevant to understand actual scientific practice nor meaningful to practising scientists or other research-concerned communities. In understanding of scientific practice, philosophical concepts benefit empirical insights—and empirical work profits from conceptual reflection. But how should, how can philosophers create such a dialogue between abstract and concrete?

We have given two examples of qualitative empirical work which addresses concepts of Philosophy of Science. These examples are drawn from our own work. In our work, we focus on actual scientific practice as performed by a community of people in reaction to each other. In doing so, we pay close attention to the procedural, dynamic character of scientific knowledge creation under ever changing conditions. Despite a similar underlying motivation, our work proceeds in two rather different ways. While Mansnerus takes the study of the concrete as a starting point, Wagenknecht engages with the abstract right from the beginning. Moreover, our research interests vary and so do our objects of empirical observation. While Wagenknecht has chosen to study collaborative scientific practice, epistemic trust and epistemic dependence from the perspective of Social Epistemology, Mansnerus has chosen to follow both the actors and the objects of scientific modelling practices. These two examples show that dialoguing as a means to advance a Philosophy of Science in Practice can take various forms. Dialoguing is not a recipe to be followed strictly, but a means to reflect upon the methodological fruits and difficulties an empirically-engaged philosophy has to offer. Its fruits, we think, outweigh its difficulties.

References

Alexa, M., Zuell, C.: Text analysis software: Commonalities, differences and limitations: The results of a review. Qual. Quant. **34**, 299–321 (2000)
Ankeny, R., Chang, H., et al.: Introduction: Philosophy of Science in Practice. Eur. J. Philos. Sci. **1** (3), 303–307 (2011)
Attride-Stirling, J.: Thematic networks: an analytic tool for qualitative research. Qual. Res. **1**, 385–405 (2001)
Auranen, K.: On Bayesian modelling of recurrent infections. Doctor of Philosophy Article, Rolf Nevanlinna Institute, University of Helsinki (1999)
Bogner, A., Menz, W.: The theory-generating expert interview: epistemological interest, forms of knowledge, interaction. In: Bogner, A., Littig, B., Menz, W. (eds.) Interviewing Experts, 43–80. Basingstoke, Palgrave (2009)
Boyatzis, R.E.: Transforming Qualitative Information Thematic Analysis and Code Development. Sage, London, New Delhi (1998)
Braun, V., Clarke, V.: Using thematic analysis in psychology. Qual. Res. Psychol. **3**, 77–101 (2006)
Buber, M.: I and Thou. Touchstone, New York (1970)
Burian, R.M.: The dilemma of case studies resolved: the virtues of using case studies in the history and philosophy of science. Perspect. Sci. **9**(4), 383–404 (2001)

Callon, M.: Struggles and Negotiations to Define What Is Problematic and What Is Not. The Socio-logic of Translation. Kluwer Academic Publishers, Dordrecht (1980)
Chang, H.: Inventing Temperature. Oxford University Press, Oxford (2004)
Chang, H.: The philosophical grammar of scientific practice. Int. Stud. Philos. Sci. **25**(3), 205–221 (2011)
Chang, H.: Beyond case-studies: History as Philosophy. In: Mauskopf, S., Schmaltz, T. (eds.) Integrating History and Philosophy of Science, 109–124. Springer, Dordrecht (2012)
Coffey, A., Atkinson, P.: Making Sense of Qualitative Data. Complementary Research Strategies. Sage, London (1996)
Collins, H.M.: History and sociology of science and history and methodology of economics. In: De Marchi, N., Blaug, M. (eds.) Appraising Economic Theories: Studies in the Methodology of Research Programs, 492–498. Edward Elgar Publishing Hunts, England (1991)
Crabtree, B.F., Miller, W.L.: A template approach to text analysis: developing and using codebooks. In: Crabtree, B.F., Miller, W.L. (eds.) Doing Qualitative Research: Multiple Strategies, 93–109. Sage, Newbury, London (1992)
Czarniawska, B.: Shadowing and other techniques for doing fieldwork in modern societies. Malmoe, København, Oslo, Liber/CBS/Universitetsforlaget (2007)
de Sousa, R.: The Rationality of Emotion. MIT Press, Cambridge, MA (1987)
Douglas, H.: Engagement for progress: applied philosophy of science in context. Synthese **177**, 317–335 (2010)
Elliott, K.C.: Epistemic and methodological iteration in scientific research. Stud. Hist. Philos. Sci. **43**, 376–382 (2012)
Ellis, C., Berger, L.: Their story/my story/our story: including the researcher's experience in interview research. In: Holstein, J.A., Gubrium, J.F. (eds.) Inside Interviewing. New Lenses, New Concerns, 467–494. Sage, London (2003)
Fagan, M.B.: Is there collective scientific knowledge? Arguments from explanation. Philos. Q. **61**(243), 247–269 (2011)
Fehr, C., Plaisance, K.S.: Socially relevant Philosophy of Science: an introduction. Synthese **177**, 301–316 (2010)
Fontana, A., Frey, J.H.: The interview: from structured questions to negotiated text. In: Denzin, N. K., Lincoln, Y.S. (eds.) Handbook of Qualitative Research, 645–672. Sage, London, New Delhi (2000)
Fox Keller, E.: A Feeling for the Organism. The life and Work of Barbara McClintock. W. H. Freeman and Company, New York (1983)
Friedman, M.S.: Martin Buber. The Life of Dialogue. University of Chicago Press, Chicago (1955)
Geertz, G.: Thick description. Toward an interpretive theory of culture. Contemporary field research. Perspectives and formulations. R. Emerson, Waveland Press, Illinois (1973/2001)
Giere, R.N.: Explaining Science. A Cognitive Approach. The University of Chicago Press, Chicago, London (1988)
Gilbert, M.: Collective belief and scientific change. In: Gilbert, M., Lanham, B. (eds.) Sociality and Responsibility: New Essays in Plural Subject Theory, 37–49. Rowman and Littlefield, New York (2000)
Hacking, I.: Representing and Intervening: Introductory Topics in the Philosophy of Natural Science. Cambridge University Press, Cambridge (1983)
Hacking, I.: The looping effects of human kinds. In: Sperber, D., Premack, D., Premack, A. (eds.) Causal Cognition: An Interdisciplinary Approach, 351–383. Oxford University Press, Oxford (1995)
Hakkarainen, K., Sintonen, M.: The interrogative model of inquiry and computer-supported collaborative learning. Sci. Educ. **11**, 25–43 (2002)
Hardwig, J.: Epistemic dependence. J. Philos. **82**(7), 335–349 (1985)
Hardwig, J.: The role of trust in knowledge. J. Philos. **88**(12), 693–708 (1991)
Hasu, M., Miettinen, R.: Dialogue and intervention in Science and Technology Studies: whose point of view? Working paper series of the Center for Activity Theory and Developmental

Work Research, **35**, 1–43 (2006). http://www.edu.helsinki.fi/activity/publications/files/333/hasu_and_miettinen_2006.pdf

King, N., Horrocks, C.: Interviews in Qualitative Research. Sage, Thousand Oaks (2010)

Kitcher, P.: The Naturalist Return. Philos. Rev. **101**, 53–114 (1992)

Knorr-Cetina, K.: The Manufacture of Knowledge: An Essay on the Constructivist and Contextual Nature of Science. Pergamon Press, Oxford (1981)

Knorr-Cetina, K.: Sociality with objects: social relations in postsocial knowledge societies. Theor. Cult. Soc. **14**(4), 1–30 (1997)

Kuhn, T.S.: Notes on Lakatos. Proceedings of the Biennial Meeting of the Philosophy of Science Association 1970, in memory of Rudolf Carnap. In: Buck, R.C., Cohen, R.S. (eds.) Reidel Publishing, vol. 8, 137–146 (1971)

Kuhn, T.S.: The halt and the blind. Br. J. Philos. Sci. **31**, 181–192 (1980)

Lakatos, I.: History of science and its rational reconstructions. Proceedings of the Biennial Meeting of the Philosophy of Science Association 1970, in memory of rudolf carnap. In: Buck, R.C., Cohen, R.S. (eds.) Reidel Publishing, vol. 8, 91–136 (1971)

Latour, B.: Pandora's Hope: Essays on the Reality of Science Studies. Harvard University Press, Cambridge, MA (1999)

Latour, B., Woolgar, S.: Laboratory Life: The Social Construction of Scientific Facts. Sage Publications Ltd., London (1979/1986)

Mattila, E.: Questions to artificial nature: a philosophical study of interdisciplinary models and their functions in scientific practice. Philosophical studies from the University of Helsinki (2006)

Mol, A.-M.: The body multiple. Ontology in medical practice. Duke University Press, Durham, London (2002)

Morgan, M.: Models, stories and the economic world. J. Econ. Methodol. **8**(3), 361–384 (2001)

Morgan, M.S., Morrison, M.: Models as Mediators: Perspectives on Natural and Social Science. Cambridge University Press, Cambridge (1999)

Nersessian, N.J.: The method to "meaning": a reply to Leplin. Philos. Sci. **58**(4), 678–686 (1991)

O'Malley, M.: Exploration, iterativity and kludging in synthetic biology. C. R. Chim. **14**, 406–412 (2011)

Osbeck, L.M., Nersessian, N.J., Malone, K. R., Newstetter, W. C.: Science as Psychology. Sense-Making and Identity in Science Practice. Cambridge University Press, Cambridge (2011)

Pitt, J.C.: The dilemma of case studies: toward a heraclitian philosophy of science. Perspect. Sci. **9**(4), 373–382 (2001)

Polanyi, M.: Personal Knowledge. Towards a Post-Critical Philosophy. University of Chicago Press, Chicago (1962)

Rehg, W., Staley, K.W.: The CDF collaboration and argumentation theory: the role of process in objective knowledge. Perspect. Sci. **16**(1), 1–25 (2008)

Schickore, J.: More thoughts on HPS: another 20 years later. Perspect. Sci. **19**(4), 453–481 (2011)

Sintonen, M.: The two aspects of method: questioning fellow inquirers and questioning nature. In: Sintonen, M. (ed.) The Socratic Tradition: Questioning as Philosophy and as Method. College Publications, UK (2004)

Sismondo, S.: Models, simulations and their objects. Sci. Context **12**(2), 247–260 (1999)

Staley, K.W.: Evidential collaborations: epistemic and pragmatic considerations in 'group belief'. Soc. Epistemol. **21**(3), 249–266 (2007)

Strauss, A., Corbin, J.: Basics of Qualitative Research Grounded Theory Procedures and Techniques. Sage, Newbury Park, London, New Delhi (1990)

Strauss, A.L.: Qualitative Analysis for Social Scientists. Cambridge University, New York, Cambridge (1987)

Thagard, P.: Computational Philosophy of Science. MIT Press, Cambridge, MA (1988)

Wagenknecht, S.: Opaque and translucent epistemic dependence in collaborative scientific practice. Episteme **11**(4), 475–492 (2014)

Wagenknecht, S.: Facing the Incompleteness of Epistemic Trust: Managing Dependence in Scientific Practice. Social Epistemology **29**(2), 160−184 (2015)
Winsberg, E.: Sanctioning models: The epistemology of simulation. Science in context **12**(2), 275−292 (1999)
Wray, B.K.: The epistemic significance of collaborative research. Philos. Sci. **69**, 150–168 (2002)
Wylie, A.: Thinking from Things. University of California Press, Berkeley, CA (2002)
Zuckerman, H.: Interviewing an ultra-elite. Public Opin. **32**(2), 159−175 (1972)

Part II
Case Studies

Modeling as a Case for the Empirical Philosophy of Science

The Benefits and Challenges of Qualitative Methods

Ekaterina Svetlova

Abstract In recent years, the emergence of a new trend in contemporary philosophy has been observed in the increasing usage of empirical research methods to conduct philosophical inquiries. Although philosophers primarily use secondary data from other disciplines or apply quantitative methods (experiments, surveys, etc.), the rise of qualitative methods (e.g., in-depth interviews, participant observations and qualitative text analysis) can also be observed. In this paper, I focus on how qualitative research methods can be applied within philosophy of science, namely within the philosophical debate on modeling. Specifically, I review my empirical investigations into the issues of model de-idealization, model justification and performativity.

Keywords Empirical philosophy of science · Qualitative methods of research · Modeling · De-idealization · Model justification · Performativity

1 Introduction

In recent years, the emergence of a new trend in contemporary philosophy has been observed in the increasing usage of empirical research methods to conduct philosophical inquiries. Prinz (2008) speaks about a "methodological revolution" in philosophy and identifies its two main paths: "empirical philosophy" and "experimental philosophy". According to Prinz's classification, empirical philosophers rely on findings from other disciplines; for example, philosophers of mind use *secondary data* from cognitive sciences and psychology to develop and analyze (historical) case studies. In contrast, experimental philosophers collect data themselves, primarily applying *quantitative methods* of empirical research (experiments, questionnaires, etc.). In addition to investigations into the nature of intuition

E. Svetlova (✉)
University of Leicester School of Management, University of Leicester,
University Road, LE1 7RH Leicester, UK
e-mail: es285@le.ac.uk

(Alexander 2012), innateness (Griffiths 2002; Griffiths et al. 2009), free will and moral responsibility (Nahmias and Murray 2010) and certain other philosophical concepts, an interesting and promising movement of "experimental philosophy of science" (Griffiths and Stotz 2008) has recently emerged and attracted attention. Its proponents conduct surveys examining scientific practice in order to enrich the philosophical understanding of scientific concepts (e.g., genes).

This classification, though plausible, implies-in my view-an overly narrow understanding of empirical philosophy, the most characteristic trait of which is reliance on all kinds of empirical methods, i.e., on secondary, quantitative and *qualitative* research. What distinguishes empirical philosophy from other disciplines that apply empirical methods (e.g., psychology or the social sciences) is the fact that the collected data are used to address genuinely *philosophical* problems; thus, empirical philosophy can be defined as a branch of philosophy in which answers to philosophical questions are informed by data that has been collected by means of empirical methods.

The workshop "The empirical philosophy of science—qualitative methods" held in Sandbjerg, Denmark, in March 2012 (whose proceedings will include this paper) was an important step in establishing this understanding of empirical philosophy as based on the application of empirical research. The workshop concentrated on the importance of qualitative methods such as in-depth interviews, participant observations of scientific practices and qualitative text analyses as means of studying scientific practices. This "qualitatively informed" philosophy of science is developing into an important branch of empirical philosophy and provides the focus of this paper. In what follows, I address the value of a qualitative empirical approach for philosophers, as well as its limitations. In particular, I elaborate upon how empirical findings can be used to develop philosophical concepts and be integrated into a philosophical framework.

To this end, I will first address the ongoing discussion on the topic of experimental philosophy. The use of empirical (in this case, quantitative) data for philosophical argumentation is often perceived as radically opposed to the traditional philosophical methodology, i.e., formal logic and conceptual analysis. Consequently, experimental philosophers must justify the application of empirical methods in their philosophical investigations. For the emerging field of the empirical philosophy of science, it might be instructive to review this debate in order to clarify proponents' position with regard to both the benefits and the limitations of an empirical approach to philosophical inquiry (Sect. 2).

In Sect. 3, I will reflect upon the area within the philosophy of science to which I apply qualitative methods of empirical research, namely *modeling*. I will show that philosophers have already recognized that investigations of modeling increasingly require a re-focusing away from the abstract theoretical issue of what models are and how they relate to theories and the world towards a practice-oriented approach, i.e., an increasing concentration on the concrete functioning of models in scientific investigations as well as in applied fields such as politics and economics. This concrete functioning can best be approached by the application of methods that allow investigation into particular human *practices*, i.e., by the application of qualitative methods of empirical research (interviews, observations, etc.). In other words, modeling represents a good example of a topic within the philosophy of science in which the usage of these methods could be especially beneficial.

In Sect. 4, I will present three examples from my research, demonstrating how the use of qualitative empirical methods allows me to address genuine philosophical questions from a new perspective. The examples illustrate three cases in which the application of qualitative methods could be especially valuable: first, where there are many theories about the same phenomenon and the discussion needs a new direction; second, where there is an a priori philosophical theory that could or should be challenged from the point of view of empirical results; and, third, where the background mechanism of a certain phenomenon is unclear. My examples relate to the issues of model de-idealization, model justification and performativity. In all of these cases, the philosophical concepts are used as a baseline to be compared with the empirical findings. In Sect. 5, I will briefly summarize the findings of the paper and discuss the major challenges of the application of empirical methods in philosophy.

2 Lessons from Experimental Philosophy

Recently, a movement has emerged within analytic philosophy whose participants have sought to challenge the traditional philosophical approach by using empirical methods that are typical of the social and cognitive sciences. This movement is discussed in the literature under the label of "experimental philosophy" (e.g., Knobe and Nichols 2008). Characteristically, experimental philosophers collect data using primarily quantitative research methods (experiments, questionnaires, etc.).

Based on their empirical findings, these scholars question, for example, the validity of philosophical intuition (e.g., Alexander 2012). Traditionally, philosophical claims are grounded in intuition, which often does not require any further evidence. There is an assumption "that our own philosophical intuitions are appropriately representative"; however, this assumption "turns out to be a *bad* habit. It ignores our tendency to overestimate the degree to which others agree with us." (ibid., 1). Experimental philosophers claim that "the favorite method of traditional philosophers—asking yourself what everyone thinks—seems hopelessly outdated" (Lackman 2006); philosophers should not guess what other people are thinking, but must instead ask what and how they think. This "asking" implies the application of empirical methods of psychology and other cognitive sciences, including controlled and systematic experiments and surveys. The goal is to study how other people (i.e., non-philosophers) make judgments about philosophical issues—for example, how they form intuitions about knowledge (Weinberg et al. 2001) or references (Machery et al. 2004). Typically, experimental philosophers construct a case, present the case to the laymen (i.e., those who are not philosophically educated) and collect the responses to their questions about the case. The resulting data allow experimental philosophers to challenge the implicit claim made by professional philosophers that their positions coincide with the views of ordinary people ("common sense"). There is often also a discrepancy between the philosophical and folk intuitions that underlie philosophical assertions, and it can be demonstrated that intuitions vary among cultures. These findings represent the experimental philosophers' challenge to traditional philosophy and should be taken into consideration.

However, these challenges and results are often neglected or ignored by "traditional" philosophers who doubt that empirical approaches can make any contribution to philosophy. Demonstrating that such a contribution is possible is the primary goal of the new experimental movement, which explains the sustained focus on the question of why empirical data are philosophically interesting. Experimental philosophers constantly stress the relevance of empirical data for philosophical inquiries: "Whereas the 'experimental' part of the name refers to the fact that they run studies and collect data concerning folk intuitions, the 'philosophy' part refers to the fact that they discuss the various implications these data have for philosophical debates" (Nadelhoffer and Nahmias 2007, 125). I believe that a convincing demonstration that there are indeed important implications of such data for philosophical debates is crucial for the viability of empirical philosophy in general. The field's proponents must establish the connection between data and theory in the resolution of genuinely philosophical problems. I agree with Prinz (2008) and Griffiths and Stotz (2008) that the difference between disciplines should not be defined by methodology (empirical evidence vs. introspection) but rather by the types of questions asked by the researchers: "Experimental philosophers have not lost their identity as philosophers through their employment of methods traditionally associated with the sciences, because they employ these methods in an attempt to answer philosophical questions" (Griffiths and Stotz 2008, 3).

For example, experimental philosophers have raised concerns about the use of intuition as the basis of philosophical practice. They have demonstrated that different people have different intuitions, and that this diversity depends on many factors, such as gender, age, ethnicity and culture (Alexander 2012, 3). However, it is important to stress that these findings do not merely contribute to the (for *psychologists*, salient) question of what determines the formation of intuitional judgment; in addition, experimental philosophers use their data to address central *philosophical* questions, such as the consequences this diverse range of intuitions may have for the formation of philosophical judgment, the determination of whose intuition is important and whose may be neglected in philosophy and, more generally, how cognition produces or influences philosophical understanding.

The answers to such philosophical questions may be identified when data are related to theories, which is why philosophers should carefully consider how they can meaningfully *combine* formal conceptual analysis and empirical results (Knobe 2007; Griffiths and Stotz 2008; Crupi and Hartmann 2010). In the case of experimental philosophy, there is a variety of stylistic options: "some experimental philosophers use data about ordinary intuition to support philosophical theories; others use such data to better understand the psychological mechanisms that generate such intuitions, while still others gather such data to show that some intuitions may be too unreliable to support philosophical theories in the first place" (Nadelhoffer and Nahmias 2007, 123). Note that the "theory-data" axis is the focus of all the projects mentioned here.

The primary benefits from empirical (experimental) approaches have thus far included contributions to theoretical concepts, the ability to challenge existing theories and the suggestion of new conceptual directions for research. Some examples of these benefits are described below.

Experiments and questionnaires have been used within experimental philosophy to test hypotheses that were formulated in a purely theoretical context. Moreover, experiments might also lead to new hypotheses; however, "these hypotheses are not put forward in a theoretical vacuum: they might relate to an existing theoretical framework, and so some tinkering may have to be done to fit the new hypotheses (or a modified version of it) into the theoretical framework (or a modified version of it). In short, experimental data may provide guidance and insight in theory-construction in a number of ways" (Crupi and Hartmann 2010, 88). For example, Crupi and Hartmann (2010) demonstrate how philosophers who use empirical data on human cognition and behavior could extend the Bayesian account of confirmation "from basic probability theory to more advanced formal notions with distinct philosophical origins" (ibid., 94). Furthermore, they consider empirical methods to be useful in situations in which there is "a spectrum of different theories" concerning one particular phenomenon (e.g., scientific explanation); in this case, "empirical studies may stir the debate in a new direction" (ibid., 93). The proponents of experimental philosophy of science have stated that empirical data on conceptual diversity within scientific communities could contribute to first-order theories on why certain scientific insights are conceptualized in one particular way and not another (Griffiths and Stotz 2008).

The empirical philosophy of science that is based on the application of *qualitative* methods could claim for itself advantages and benefits similar to those afforded to the field of experimental philosophy. As with experimental philosophy, the "qualitatively oriented" empirical philosophy of science could reveal the discrepancy between *philosophical claims* (which are based solely on intuitive abstraction) and the real practice of knowledge production, as discovered by means of empirical methods. I am convinced that the major contribution of empiricism to philosophy in general lies in its ability to draw attention to the inconsistencies between introspective conceptual analysis and concrete empirical examples, as well as to take such inconsistencies into account theoretically.

The particular advantage of the application of qualitative empirical methods of research, however, is that it allows data—not theory—to take the lead. Usually, philosophical investigations are led by an elaborated argument that is illustrated (confirmed or challenged) by examples from other sciences, or, as in the case of experimental philosophy, quantitative research is conducted that is confirmatory in nature. The central advantage of qualitative methods is their explorative character: Because they are not generally used to test hypotheses that are derived from theory, they are able to produce new insights about phenomena and generate new knowledge (Flick 2009; Silverman 2010). Qualitative methods allow investigation into the deeper background issues of phenomena that form part of human practices, for example, the various scientific practices of knowledge production. Furthermore, they require that the data speak their own language and are accepted in their own right—not as the confirmation of an argument but as the ultimate focus and point of departure of the (theoretical) inquiry. Even though qualitative methods do not always generate new theories, they actively participate in the development of theories: Situational research that focuses on agents' interpretations permits richer conceptual possibilities and is able to question existing theories more profoundly than quantitative methods can. Thus, the application of *qualitative* methods may leverage the advantages of case-study philosophy as well as those of experimental philosophy, a field that is primarily committed to quantitative methods.

3 Modeling Practice as a Prime Case for the Application of Qualitative Methods

The field within the philosophy of science to which I apply qualitative methods—modeling—could be used as a prime example for the discussion of the benefits of a qualitative empirical approach for philosophers. This is so because the recent philosophical debate on modeling has recognized the need for a deep understanding of scientific practices, e.g., the practices of model creation and model use; to achieve this understanding, qualitative empirical methods could be of particular benefit. The traditional method of epistemology as an a priori, purely analytic investigation has more recently been questioned. More concretely, in addition to the established syntactic and semantic views on models, the practice-oriented focus on the roles and functioning of models in science has slowly but surely crystallized (Morgan and Morrison 1999; Knuuttila et al. 2006). It is significant to note that this theoretical movement has been institutionalized by the Society for Philosophy of Science in Practice.[1] According to its research program, models should be studied as elements of scientific practice and thus the thorough investigation into how models are used and how they function within this practice is crucial for understanding the nature of models, their roles and how they produce explanations or represent phenomena.

Knuuttila (2005a, 2011) attacks the understanding of models as pure representational structures and takes the practice-oriented approach as a point of departure. Her studies demonstrate how an established philosophical stance toward modeling, i.e., representation, can be challenged by, among others things, empirical insights. She argues that models are epistemic artifacts or tools that are purposely created for particular practical goals and are *made* productive by means of human intervention and manipulation within particular scientific practices. The definition of models as epistemic tools situates them as material objects that are not "ready-made" but

---

[1] www.philosophy-science-practice.org.

rather unfolding elements of situational practices (Knuuttila and Merz 2009; Knorr Cetina 1997, 2001; Rheinberger 1997). This theoretical move "means also leaving the conceptual and ideal world of philosophy and entering into the social and material world of human actors, where material objects, usually human-made artefacts, draw together numerous activities and different actors" (Knuuttila 2005b, 48)—the typical field for the application of qualitative research methods.

Within this practice-oriented debate, there is an increased focus on the pragmatic aspects of model use, which allows for the explanation of what makes models useful tools despite their generic character, their inaccuracy, and their tenuous connections with the real world (Morgan and Morrison 1999; Mäki 2009). Attention is paid not merely to models as such (their structure, means and forms of idealization, etc.) but-again-to modeling practices and their contexts: to comprehend the very nature of models, we must take into consideration the analysis of additional factors such as the role of model users and their prospective purposes and narratives. It is important to keep in mind that qualitative methods are designed specifically for the study of human actions with due regard to their specific context.

The natural consequences of this new conceptualization are an increased interest in the material practice of model construction and manipulation and the empirical aspects of this interest. In this context, a clear connection to science and technology studies (STS) can be observed: "…the studies of models by philosophers and STS scholars can be seen to interact with, intersect and complement one another, with the practice-orientation laying out a bridge between the two" (Knuuttila et al. 2006, 4f.). It is important to stress that the philosophy of science and STS do not just share content; there is also a methodological exchange that occurs when the qualitative methods of empirical research, which are typical for STS, gain a stronger hold of the philosophy of science due to the reasons discussed above.

Knuuttila (2005b, 19) asks: "Provided that we accept the results of empirical science as part of philosophical reasoning, should we then stop at that? Is there a place for empirical study in philosophical argumentation? I think that there is, if only because a lot of research done in the philosophy of science proceeds by presenting cases from specific disciplines, taking historical data into account as well. Since I approach representation and modeling from the point of view of scientific practice, I have felt a need to get some grasp of the practices themselves." Thus, the adaptation of qualitative empirical methods allows for the extension of the philosopher's methodological repertoire beyond the customary historical examples, and it is appropriate to analyze relevant human actions (e.g., the process of model use), the tacit knowledge of practice participants (e.g., of model users, audience, and model creators) and, particularly, the practice-specific nuances of objects (i.e. models) applications. Qualitative methods are a part of a methodology that enriches philosophical reasoning through the detailed and substantial study of actual scientific practices.

In line with this argument, Alexandrova (2008) also explicitly makes a case for the development of the "practice-based philosophy of science" (p. 384) and implicitly for the necessity of empirical investigations within philosophical inquiries on modeling. Like Knuuttila, she stresses that models are productive tools not simply because of their nature, but rather that particular efforts are required to make models work and count. Alexandrova develops a case study of a spectrum auction institutional design that is based on the standard models of game theory. Although she does not specifically apply empirical methods, she uses materials that are based on "numerous observations" of the design process (p. 391). This "practice-based" argumentation allows her to demonstrate the insufficiency of existing accounts of model application, i.e., the satisfaction of assumptions by Hausman and the capacity account by Cartwright and to develop her own account. She shows how—*in the practice* of auction design—theoretical models serve as open formulae: they inform the process of auction design while they deliver "suggestions for developing causal hypotheses that can be tested by experiments" (p. 396). This specific function of models could not have been discovered by pure analytical reasoning, without analyzing the practice of model use.

Alexandrova's work implicitly suggests the necessity for accurate empirical methods for philosophical investigations in many places: for example, she claims that interactive holism (the interwovenness of causes in economic life) might "be an empirical issue to be selected by looking at economic reality" (p. 392). Additionally, the success of models and scientific progress should be more generally grasped as an empirical context-depending issue (also Alexandrova and Northcott 2009).

To summarize, models are no longer considered by philosophers to be purely theoretical and abstract entities but rather "dirty" and insecure tools that must be manipulated and "made to count" in situ to produce knowledge; thus, their investigation requires methods that produce insights into the very practice of model creation and model use.

To complete the discussion about the relevance of qualitative methods for the philosophical debate about modeling, it is important to note that as models have increasingly been incorporated into decision-making and regulatory processes in a large variety of applied fields (e.g., politics, economy, particularly financial markets), philosophers are forced to pay attention to the application of models in a number of *non-scientific* practices. Here, the "dirtiness" and materiality of models and their interwovenness with the pragmatic aspects of practice is even more significant. Thus, if models are no longer analyzed as merely instruments of scientific inquiry, philosophers' a priori analytical knowledge of how science works in general might be argued to be particularly insufficient. If we perceive models as instruments of guiding fateful decisions in flood management, climate science, health care, financial markets, etc., the qualitative empirical examination of the use of models in the different practical arenas of decision-making will become especially necessary. This is what the most recent, relevant studies suggest (e.g., van Egmond and Zeiss 2010; Lane et al. 2011; Gramelsberger and Mansnerus 2012; Svetlova and Dirksen 2014). These studies radically refrain from approaching modeling as a purely scientific endeavor and emphasize that the traditional separation of science—as a place of model construction and development—from the realm of pragmatic model application by practitioners has lessened. Many models are no longer created in the "ivory tower" of science and then transferred as fixed objects to practical fields in which they are mechanically applied. Rather, recent research on modeling demonstrates that, in many cases, the "scientific life" of models cannot be separated from their "working life" (the term of Erika Mansnerus) external to science-scientific and practical criteria and interests are entwined. This means that scientific aspects may derive from this "working life" or that non-scientific fields—through their involvement in the creation and application of models—become grounded in scientific modeling as a result of which models influence political and economical decisions. Financial models are case in point for this development.

4 Empirical Examples

At this point I would like to demonstrate with concrete examples from my research some of the ways in which qualitative empirical methods can contribute to the philosophy of science. The examples that follow are from my research on the use of modeling in the field of finance.

My studies are based on research that was conducted in several German and Swiss asset management companies and banks. They consist of twenty-eight guided interviews with investment professionals. Most of the interviews took place in person, and only one was conducted by telephone. All of the interviews were recorded and transcribed. The evaluation included coding and categorizing (Corbin and Strauss 2008; Flick 2009; Silverman 2010).

Formal interviews were complemented by a three-month process of participant observation conducted in the portfolio management department of a private Swiss investment bank in Zurich. The application of financial valuation models (e.g., CAPM, BSM and DCF) was of particular interest during the course of the empirical study.

As indicated in the discussion above, empirical methods could be useful in the following cases:

- *Case (1): If many theories about the same phenomenon exist.* This case applies, for example, to the debate on model idealization and de-idealization. Based on my empirical research about the discounted cash flow (DCF) model, I demonstrate that existing accounts of de-idealization do not apply, especially when we are concerned with the not purely epistemic but with the more pragmatic practice of model use, specifically the application of models in financial markets. I show that the pragmatic aspects of model use, such as audience, market context and narrative, play the most prominent role for the analysis in such cases. Based on my field materials, I propose the concept of "de-idealization by commentary," which is supposed to be an enrichment of the existing theoretical concepts of de-idealization aimed at steering the whole debate in a more pragmatically oriented direction. Note that my conceptualization arises solely from work with empirical materials.

- *Case (2): If there is an* a priori *philosophical theory that could or should be challenged from the point of view of empirical results.* In my example, I challenge Boumans' idea of model justification (i.e., the "built-in" mechanism of model justification) and propose the "built-out" mechanism, which applies in the case of the DCF model.
- *Case (3): If the background mechanism of a particular phenomenon is not clear.* Here, I bring an example that refers to the issue of performativity—a broadly discussed concept in philosophy and STS that postulates that knowledge, theories, and models not only represent the world but also influence or constitute that which is represented. However, the performativity thesis remains vague as it does not provide a detailed conceptualization and description of how models create or change reality. Empirical investigations help to clarify which forms of performativity can be found in the practice of financial markets and how exactly financial models influence markets.

Before I go deeper into my examples, I would like to highlight the additional function of empirical investigations in philosophy that results from my research. Empirical studies can expand a philosophical framework by bringing into play new examples and opening new fields. As the Society for the Philosophy of Science in Practice formulates, "[o]ur views of scientific practice must not be distorted by lopsided attention to certain areas of science. The traditional focus on fundamental physics, as well as the more recent focus on certain areas of biology, will be supplemented by attention to other fields such as economics and other social/human sciences, the engineering sciences, and the medical sciences, as well as relatively neglected areas within biology, physics, and other physical sciences". In my case, I expand philosophical investigations into the area of financial valuation models. These models have not yet been analyzed by philosophers of science.

As mentioned above, financial modeling delivers an interesting example of the field in which the tight entwinement of the academic efforts and the context of application is especially distinct. This kind of modeling surpasses the pure "doing science"; however, this fact does not justify neglect of financial models by the contemporary philosophy of science. Financial models can be considered to be scientific objects that unfold and acquire different meanings through the various phases of their biographies, including construction, application, further development and, perhaps, later non-existence. Financial models may be developed in fundamental science and then travel to the field of their application in financial markets, or they may be constructed in the practical field of investment banking and then move to fundamental science to be further developed. Here, we are concerned with peculiar practices where scientific knowledge is produced and used in a specific way. Though many phases of the financial models' biographies take place in the academia, the focus is on models' use for the not purely scientific inquiry. Unlike scientists, financial market participants not only look for good descriptions, explanations or predictions of real-world phenomena, but also seek out and develop models that enable them to know how to act in every particular market situation—that is, how to gain positive investment returns and how to manage risks. Those practices as a specific way of doing and using science deserve an attention of philosophers of science. The pragmatic context of the models' application (actors, their goals and their practices) within financial practices differs greatly from traditional scientific settings and plays the more prominent role. Thus, financial modeling is a field where empirical methods could be of particular help.

Below, I return to my examples through which I outline exactly how qualitative methods can be beneficial for philosophical discussions.

Case (1): De-idealization by commentary

In the first empirical case study (Svetlova 2013), I discuss how a popular valuation model (the discounted cash flow model) idealizes reality and how the market participants de-idealize it in market practice. I contrast the existing accounts of model de-idealization (the relaxing of simplifying assumptions by Hausman (1992) and McMullin (1985) as well as the concretization or re-addition of the excluded unessential properties by Nowak (1980, 1989) and Cartwright (1989) with an in-depth empirical description of how the market participants de-idealize the DCF model in concrete market situations. The empirical research demonstrates a discrepancy between established philosophical accounts and what we find in the markets.

In contrast to Cartwright's view that economic models are generally over-constrained (Cartwright 1999, 2009), I suggest that valuation models are under-constrained. Although, at first glance, the DCF model is based on a theoretically valid causal mechanism that contains just two main factors (i.e., future cash flows and the discount rate), one can demonstrate that those determinant parameters are non-observable and vague and that they depend on the calculation of additional parameters (future sales, growth rates, profit margins, capital expenditures, assumptions about investments, including working capital and fixed investment as well as some macro parameters; this list is not exhaustive). Thus, the DCF model is not based on a narrow clear structure (i.e., it is not over-constrained), and it is not perfectly idealized; rather, the model is too rich and loose.

This observation serves as the reason why, in the financial markets, neither the relaxation of assumptions nor concretization is the prevailing method of de-idealization. It is not a problem of the omission of many relevant causes that should be added back; on the contrary, too many factors are implicitly included in the model. Thus, the introduction of more realistic assumptions in the form of adding back or the making explicit of further factors would increase the model complexity and fail to provide a bridge between the model and the world. In the case of under-constrained models, the implementation of additional factors misses the point. How then are the under-constrained financial models de-idealized and used?

By answering this question, the power of qualitative empirical methods can be seen: they provide an insight into how de-idealization *happens*. Interviews with model users as well as observations of particular examples of model application show that every user specifies his or her own DCF model; i.e., he or she determines the definitive structure and parameters and, hence, completes the process of idealization in situ. Surprisingly, the use of empirical methods also demonstrates that valuation models are not so actively manipulated and changed as philosophical accounts suggest in the case of scientific models. Rather, once the model has been finalized, the users in financial markets prefer to keep their individual model version stable and avoid the constant changing and adjustment of parameters. They arrive at investment decisions while they compare model results with their own feelings or judgments concerning asset classes or companies. If there is no fit between the numerical model outputs and the investors' qualitative views, then the subjectively perceived inadequacies of the model are corrected in situ, or, as market participants say, they are "overlaid": decisions are guided by investors' views rather than by formal models. In my research, I adopted the "native" empirical term and described the whole process of model de-idealization as "qualitative overlay."

These empirical findings reveal the necessity of accounting for the discrepancy between pre-established philosophical views and the realities of markets. I suggest directing the theoretical work toward the already discussed pragmatics of model use, specifically highlighting the *empirically verified* importance of story-telling as an external factor of model adjustment. Thus, I focus on de-idealization through the commentary of users. Using my empirical materials, I demonstrate how portfolio managers use narrative as a vehicle to express their holistic judgments about the market, the asset class or the company and how those judgments are formed. Narratives include all of the factors and dynamics that have been excluded, not specified or merely implied by the model in the process of decision-making; in this sense, judgment is the instrument of de-idealization.

To summarize, the empirical investigations in this case produced an interesting example of a de-idealization pattern that does not fit with the existing philosophical accounts and even allowed for the development—out of the empirical materials—of a proposal for an alternative account. This account, as suggested in the general discussion about methods above, is not isolated but is rooted in and connected to the existing theoretical (specifically, pragmatic) account of modeling. Furthermore, the empirical findings suggest that there is no unique way to de-idealize models; i.e., there are many possible ways to reduce the distortion between models and reality depending on *the style of model use*. Thus, it would be beneficial to search for further styles of de-idealization and to investigate them empirically.

At the end of the study, I also empirically constructed the hypothesis that the more under-constrained the model is, the larger the role that narrative and other pragmatic elements outside of the model play when the model is applied. This hypothesis is also one of the results of empirical research, and it should be examined on the basis of further case studies from both economic theory and practice.

Case (2): The "built-out" mechanism of model justification

To provide another example of the advantages of empirical methods for the practice of philosophy, I would like to focus on the issue of model justification. I continue with my empirical case study on DCF and ask the following: if the traditional account of de-idealization does not apply, how can we justify the use of valuation models? Can the whole issue be reduced to the application of narrative?

The issue of model justification again allows for the discussion of the discrepancy between the pure philosophy-of-science position and the empirical view. The detailed empirical description of the application of valuation models notes the differences in the justification mechanisms of the purely scientific models on the one hand and the financial models on the other.

The pragmatic accounts of models—e.g., models as "open formulae" or "raw materials" (Alexandrova 2008, 2009), as "epistemic objects" (Boon and Knuuttila 2009; Knuuttila 2011), as "mediators" (Morrison and Morgan 1999) or "boundary objects" (Star and Griesemer 1989)—suggest that there are some useful approaches to the justification of model use in cases where the traditional concepts of idealization and representation do not apply. However, all of the pragmatic accounts mentioned here focus on the *epistemic* function of models. They concentrate on the practice of scientific inquiry and investigate how scientists construct or manipulate models to create institutional design (Alexandrova 2008), draw inferences and reason (Boon and Knuuttila 2009) and provide understanding between various scientific communities (Star and Griesemer 1989). However, because de-idealization takes different forms in financial markets than it does in the scientific context, the justification of model use needs to be analyzed differently; the precise ways in which this process occurs can be determined empirically.

The justification of financial models is not based on a "built-in" mechanism, as is often the case for scientific models (Boumans 1999). Scientific model-builders constantly include elements of theory, data, tacit knowledge and experience directly into the model so that "a trial and error" process goes on "until all the ingredients, including the empirical facts, are integrated" (Boumans 1999, 95; van Egmond and Zeiss 2010, 65). In the case of financial valuation models, there is no such process because the models are, as described above, kept stable. The role of the "built-in" mechanism is undertaken by the ongoing commentary that takes place "outside" of the model and provides for the necessary adjustments to a continually changing, complex world.

This "outside-of-model" adjustment mechanism facilitates investment decisions on the one hand and determines the important but still subordinate role that valuation models play in decision-making on the other. This observation stresses the intermediateness of model influence on markets. Models do not enable decisions by indicating what the correct valuation of an asset is. Market participants often stressed in interviews that they do not trust models' calculations. As models are constantly overruled in the "outside" process of judgment application, they do not entirely determine the success or failure of decisions. The success of a model and the success of a decision are two different things. The justification of model use lies not in their facilitating correct decisions but in providing guidance in structuring the decision making process; the fulfillment of this function could make even a flawed model successful.

Case (3): The mechanism of performativity

Using empirical methods, I also investigated the mechanism of performativity (Svetlova 2012; Svetlova and Dirksen 2014). The performativity thesis has its roots in philosophy (Austin 1962; Derrida 1988); it has been, however, adapted and further developed by social scientists studying finance and economics (Callon 1998; MacKenzie 2003, 2006), and recently it was echoed again in some philosophical inquiries on modeling (e.g., Mäki 2011; Knuuttila 2005b).

Performativity is a slippery concept because its mechanism cannot merely be grasped analytically. First of all, it remains unclear if performativity implies that a new phenomenon (a social fact like marriage or market) is created in the process of speaking ("Austian performativity," and later, "Barnesian performativity" in the work of MacKenzie (2006)) or whether reality is merely influenced or changed by any kind of speech, theory or model ("generic" or "effective" performativity by MacKenzie (2006)). Furthermore, one finds only vague indications in the literature concerning the question of *how* exactly a speech act or a model (understood as an utterance) create a new state of affairs, i.e., new social facts, in the very moment of utterance or through which channels the influence of, for example, financial models on markets takes place. What is the exact mechanism behind such influence?

Here, again, empirical methods could provide some useful insights. Semi-structured interviews and participant observations demonstrated that there is no evidence of strong (Austian or Barnesian) performativity in the markets; however, the empirical materials do support notions of generic and also in part effective performativity. The explanation for those findings could be provided through deeper empirical investigations into the practice of model use.

An extensive and rigorous use seems to be the essential pre-condition for a model to become performative: MacKenzie's example is most often the Black Scholes option pricing model (BSM), which (at least in a period after the introduction of the model) had the effect that the real market prices came to approximate the calculated prices (MacKenzie 2003, 2006). This happened because market participants applied the Black schools model more and more as the basis for their market positions. MacKenzie and Millo (2003, p. 123) describe how the BSM became "a guide to trading": initial doubts and concerns about the model were overcome so that traders started to believe in the model and to use it to calculate option prices. Thus, the practice of model use is also, in the case of performativity, the crux for understanding how models work and influence reality, and this fact again justifies a commitment to qualitative empirical methods.

An extensive empirical study on the application of various valuation models (e.g., the DCF model, the capital asset pricing model and some option trading models) demonstrated that most examples reveal models' indirect use as described in the DCF case study in the sub-sections *Case (1)* and *Case (2)*: investors' judgments overlay the model results which—through this process—can become irrelevant for the decisions. Models, even if applied to value assets, do not have a chance to directly influence markets: as the model results are not strictly incorporated into decisions and, hence, do not enter the market, models' "utterances" stay irrelevant for what happens in the market. Hence, the actual ways in which models are used in practice prevent them from shaping reality.

Thus, the assumption of the performativity thesis that the relationship between models and reality is straight and direct; i.e., that if models are used, they immediately create or change reality, contradicts the empirical findings. Empirical investigations suggest that the relationship between models and reality is rather strongly mediated by use in *social context*: the impact of models on the economy is framed by institutional and organizational settings (e.g., the institutionalized decision-making process, the structure of departments, institutional culture with respect to the trust or mistrust of models, etc.). The social context determines whether models are strongly or just "generically" or "effectively" performative. In some institutions, the performative power of models is obviously limited in the process of their application; in some others, models have a more direct and strong influence and, thus, more power to influence markets. I deliver a detailed empirical description of this mechanism for the case of the wealth management department in a large bank in Zurich. Again, I could not have come to those insights about the importance of the mediated institutional context for understanding the performativity mechanism by way of a purely analytical methodology.

5 Conclusion and Open Questions

In this paper, I delivered arguments for the use of qualitative empirical methods in philosophical research and discussed examples to demonstrate how those methods can provide philosophical insights. I showed that qualitative methods are particularly useful if their application aims to contribute to philosophical concepts that are related to practices of any kind, to ways of how people do things. Scientific practices in general and modeling as a particular way of knowledge production are concrete powerful cases in point. Hence, qualitative methods should further be promoted to become established instruments of empirical philosophy of science.

Over the course of writing this paper, however, I noticed how many open questions and unsolved problems still confront any philosophers who choose to commit to the application of qualitative methods. Though the detailed methodological discussion would go far beyond the scope of this paper, I would like to highlight the following problems.

In addition to the rather traditional methodological questions of how to cope with subjectivity and one-case orientation of qualitative methods, empirically oriented philosophers need to address *the peculiarities of methods use in philosophy*: How should an empirical project be designed to facilitate the collection of philosophically relevant data? Are there any peculiarities concerning data analysis? In other words, what are the specificities of a *philosophical* empirical project? I think that—though qualitative methods are explorative in nature—their application in philosophy should be strongly theory-oriented (my examples in the paper support this view). It means that interview guidelines and participant observation concepts should result from a thorough analysis of philosophical texts and intensive work with philosophical notions of topics in question. This was the case in examples 1 and 3 in this paper: the concepts of de-idealization and performativity guided the empirical work. At the same time, the importance of question how models' results are justified appeared to me during the evaluation of the data; however, this prompted me to carefully read the relevant philosophical texts on model justification, to identify discrepancies between my empirical findings and philosophical theory and to reflect on those discrepancies. Thus, in contrast to social science where pure, empirically based description is sometimes accepted as an investigation result, philosophical empirical projects are much stronger related to or guided by the theoretical considerations. Still, the question of how exactly one can design a methodologically correct empirical project in philosophy is, in my view, open. This question though needs to be answered carefully by philosophers who are convinced of the benefits of qualitative empirical methods to maintain a voice in the general philosophical discussion. The workshop in Sanbjerg was obviously just the first step of a long journey.

References

Alexander, J.: Experimental Philosophy: An Introduction. Polity Press, Cambridge/Malden (2012)
Alexandrova, A.: Making models count. Philos. Sci. **75**, 383–404 (2008)
Alexandrova, A., Northcott, R.: Progress in Economics: Lessons from the Spectrum Auctions. In: Kincaid, H., Ross, D. (eds.) The Oxford Handbook for Philosophy of Economics, pp. 306–337. Oxford University Press, Oxford (2009)
Austin, J.: How to Do Things with Words: The William James Lectures Delivered at Harvard University in 1955. Oxford University Press, Oxford (1962)
Boon, M., Knuuttila, T.: Models as Epistemic Tools in Engineering Sciences: A Pragmatic Approach. In: Meijers, A. (ed.) Handbook of the Philosophy of Sciences: Philosophy of Technology and Engineering Sciences, pp. 693–726. Elsevier Science, North Holland (2009)
Boumans, M.J.: Built-in justification. In: Morrison, M., Morgan, M.S. (eds.) Models As Mediators, pp. 66–96. Cambridge University Press, Cambridge (1999)
Callon, M.: Introduction: The Embeddedness of Economic Markets in Economics. In: Callon, M. (ed.) The Laws of the Market, pp. 1–57. Blackwell, Oxford und Madlen (1998)
Cartwright, N.: Nature's Capacities and Their Measurement. Oxford University Press, Oxford (1989)
Cartwright, N.: The Dappled World: A Study of the Boundaries of Science. Cambridge University Press, Cambridge (1999)
Cartwright, N.: If No Capacities Then No Credible Worlds. But Can Models Reveal Capacities? Erkenntnis **70**(1), 45–58 (2009)
Corbin, J.M., Strauss, A.L.: Basics of Qualitative Research: Techniques and Procedures for Developing Grounded Theory. Sage Publications, Los Angeles (2008)
Crupi, V., Hartmann, S.: Formal and Empirical Methods in Philosophy of Science. In: Stadler, F. (ed.) The Present Situation in the Philosophy of Science, pp. 87–98. Springer, Dordrecht (2010)
Derrida, J.: Signature Event Context. In: Limited Inc., pp. 1–24. Northwestern University Press, Evanston (1988)
Flick, U.: An Introduction to Qualitative Research. Sage, London (2009)
Gramelsberger, G., Mansnerus, E.: The inner world of models and Its Epistemic Diversity: Case of Climate and Infectious Disease Modelling. In: Bissell, C., Dillon, C. (eds.) Ways of Thinking, Ways of Seeing, pp. 167–196. Springer, Berlin/Heidelberg (2012)
Griffiths, P., Stotz, K.: Experimental philosophy of science. Philos. Compass **3**, 507–521 (2008)
Griffiths, P., Machery, E., Linquist, S.: The Vernacular Concept of Innateness. Mind Lang. **24**(5), 605–630 (2009)

Grifiths, P.: What is Innateness? Monist **85**, 70–85 (2002)

Hausman, D.: The inexact and separate science of economics. Cambridge University Press, Cambridge (1992)

Knorr Cetina, K.: Sociality with objects. Social relations in postsocial knowledge societies. Theory Culture Soc. **14**, 1–30 (1997)

Knorr Cetina, K.: Objectual practice. In: Schatzki, T.R., Knorr Cetina, K., von Savigny, E. (eds.), The Practice Turn in Contemporary Theory, pp. 175–188. Routledge, London/New York (2001)

Knobe, J.: Experimental philosophy. Philos. Compass **2**(1), 81–92 (2007)

Knobe, J., Nichols, S. (eds.): Experimental Philosophy. Oxford University Press, Oxford (2008)

Knuuttila, T.: Models, representation, and mediation. Philos. Sci. **72**, 1260–1271 (2005a)

Knuuttila, T.: Models as epistemic artefacts: toward a non-representationalist account of scientific representation. Academic Dissertation, University of Helsinki, Faculty of Arts, Department of Philosophy and Faculty of Behavioural Sciences, Department of Education, Philosophical Studies from University of Helsinki 8. http://ethesis.helsinki.fi/julkaisut/hum/filos/vk/knuuttila/modelsas.pdf. (2005b)

Knuuttila, T.: Modeling and representing: an artefactual approach. Stud. Hist. Philos. Sci. **42**, 262–271 (2011)

Knuuttila, T., Merz, M.: Understanding by modeling: an objectual approach. In: de Regt, H.W., Leonelli, S., Eigner, K. (eds.) Scientific Understanding: Philosophical Perspectives, pp. 146–168. University of Pittsburgh Press, Pittsburgh (2009)

Knuuttila, T., Merz, M., Mattila, E.: Computer models and simulations in scientific practice. Introduction to special issue, edited by Tarja Knuuttila, Martina Merz and Erika Mattila. Sci. Stud. **19**(1), 3–11 (2006)

Lackman, J.: The X-philes: Philosophy meets the real world. Slate Magazin, http://www.slate.com/articles/health_and_science/science/2006/03/the_xphiles.html (2006)

Lane, S.N., Landström, C., Whatmore, S.J.: Imagining flood futures: risk assessment and management in practice. Philos. Trans. Royal Soc. A **369**, 1784–1806 (2011)

Machery, E., Mallon, S., Nichols, S., Stich, S.: Semantics, cross-cultural style. Cognition **92**, B1–12 (2004)

MacKenzie, D.: An equation and its worlds: Bricolage, exemplars, disunity and performativity in financial economics. Soc. Stud. Sci. **33**, 831–868 (2003)

MacKenzie, D.: An engine, not a camera: How financial models shape markets. MIT Press, Cambridge (2006)

MacKenzie, D., Millo, Y.: Constructing a market, performing theory: The historical sociology of a financial derivatives exchange. Am. J. Sociol. **109**(1), 107–145 (2003)

Mäki, U.: MISSing the World. Models as isolations and credible surrogate systems. Erkenntnis **70**(1), 29–43 (2009)

Mäki, U.: Economics making markets is not performativity. Paper presented at the XII April International Academic Conference on Economic and Social Development, the High School of Economics, Moscow, Russia, http://www.hse.ru/data/2011/04/05/1211693735/PerformativityMoscow2011.pdf (2011)

McMullin, E.: Galilean Idealization. Stud. Hist. Philos. Sci. **16**, 247–273 (1985)

Morgan, M.S., Morrison, M. (eds.): Models as Mediators: Perspectives on Natural and Social Science. Cambridge University Press, Cambridge (1999)

Nadelhoffer, T., Nahmias, E.: The past and future of experimental philosophy. Philos. Explor. **10**(2), 123–149 (2007)

Nahmias, E., Murray, D.: Experimental philosophy on free will: an error theory for incompatibilist intuitions. In: Aguilar, J., Buckareff, A., Frankish, K. (eds.) New Waves in Philosophy of Action, pp. 189–215. Palgrave-Macmillan, London, New York (2010)

Nowak, L.: The Structure of Idealization: Towards a Systematic Interpretation of the Marxian Idea of Science. Reidel, Dordrecht/Boston/London (1980)

Nowak, L.: On the (Idealizational) Structure of Economic Theories. Erkenntnis **30**, 225–246 (1989)

Prinz, J.: Empirical philosophy and experimental philosophy. In: Knobe, J., Nichols, S. (eds.) Experimental Philosophy, pp. 189–208. Oxford University Press, Oxford (2008)

Rheinberger, H.-J.: Toward a history of epistemic things: synthesizing proteins in the test tube. Stanford University Press, Stanford (1997)

Silverman, D.: Doing Qualitative Research: A Practical Handbook. Sage, London (2010)

Star, S.L., Griesemer, J.R.: Institutional ecology, 'Translations' and boundary objects: amateurs and professionals in Berkeley's Museum of Vertebrate Zoology, 1907–39. Soc. Stud. Sci. **19**(4), 387–420 (1989)

Svetlova, E.: On the performative power of financial models. Econ. Soc. **41**, 418–434 (2012)

Svetlova, E.: De-idealization by commentary: the case of financial valuation models. Synth. Int. J. Epistemology Methodol. Philos. Sci. **190**, 321–337 (2013)

Svetlova, E., Dirksen, V.: Models at Work—Models in Decision-Making. Introduction to the special section "Models at Work". Sci. Context **24**(4), 561–577 (2014)

van Egmond, S., Zeiss, R.: Modeling for policy: science-based models as performative boundary objects for Dutch policy making. Sci. Stud. **23**(1), 58–78 (2010)

Weinberg, J., Nichols, S., Stich, S.: *Normativity* and Epistemic Intuition. Philos. Top. **29**, 429–460 (2001)

Reductionism as an Identity Marker in Popular Science

Hauke Riesch

Abstract This paper takes a look at how reductionism is represented by popular science authors who have engaged in the disputes variously labelled the sociobiology, evolutionary psychology or Nature/Nurture debates. It shows how reductionism has become an identity marker through which authors on either side of the dispute signal adherence to a wider social identity, and that the philosophical content of what reductionism means gets reinterpreted according to which side of the debate the author stands on. This raises questions about the necessity to include insights from sociological theory when philosophical studies aim to include qualitative evidence on scientists' thinking.

Keywords Reductionism · Popular science · Social identity

1 Introduction

What use is philosophy to scientists? It is an often repeated assertion, following a possibly apocryphal remark attributed to Richard Feymann, that philosophy of science is as much use to scientists as ornithology is to birds. This admittedly not entirely serious remark is of course contested because, as even Feynman's own popular science writing shows, there is a deep underlying concern over scientific method and other philosophical matters within science, and there has been a recent trend within philosophy of science to philosophize more on matters that are directly relevant to scientific practice, for example, Chang's (2004) new conceptualisation of HPS (History and Philosophy of Science) as "complementary science".

H. Riesch (✉)
Department of Social Sciences, Media and Communications, Brunel University London, Uxbridge, UK
e-mail: hauke.riesch@brunel.ac.uk

Complementary to that there is a current drive to ground philosophical theorising in actual evidence on what scientists do and think. This has of course been a staple in HPS research which seeks to ground philosophy on historical evidence, and a similar trend is currently seeking to look towards social science evidence, for example Bailer-Jones' (2003) research on scientists opinions on models. As Bailer-Jones argues, the reasons for philosophers wanting to understand scientists' thinking is that there are a diversity of opinions on modelling, and instead of using scientists' opinions to construct a philosophy, she argues that it is useful for "gaining orientation", and that there is a "methodological requirement that philosophical stances towards models match the use of the term 'model' as used by scientists" (p. 276). However, I would argue that philosophical ambitions needn't end there, and that scientists' own philosophical observations and opinions can be a useful resource for the philosopher in constructing new ways of understanding their topic.

My own empirical investigations show that scientists' discourses are full of philosophical remarks and concerns, and in a previous paper (Riesch 2010a). I have used scientists' talk about a philosophical topic to draw out some possible lessons for philosophy. This paper will be a follow-up of sorts which will flag up not so much the content of scientists' philosophical thoughts, but also the social and rhetorical uses it is being put to. This points towards some of the interpretative problems inherent in using (qualitative) sociological methods in philosophy by considering the rhetorical use of a philosophical concept in scientists' popular writing, and how this potentially impacts any philosophical interpretation of scientists' philosophical writings: I will argue that use of sociological evidence in philosophy must keep in mind the critical perspective offered and developed within sociological research, because philosophical discourse can have social functions that shape and influence how they are being thought about and interpreted.

The paper will begin by outlining the theoretical background I will use to analyse the authors' representations of reductionism. I will then give a very brief account of reductionism; I will not go into a philosophical argument over what definition of reductionism should be used or under which circumstances reductionism is or is not a good idea, preferring instead to give a brief (and non-exhaustive) outline of the different usages of the term and point to some of the different definitions with which it has been used in philosophy. I will then introduce my rhetorical study of popular science: popular science is arguably the easiest genre of science writing with which to analyse scientists' rhetoric because it has a clear purpose to persuade the reader to the scientist's point of view away from the strict formal guidelines that restrict creative language use in professional scientific communication, and therefore allows the writer greater leeway to draw on a larger array of images, metaphors and philosophical concepts to advance the argument. I will then offer an interpretation of why reductionism is talked about in popular science as it is, by drawing on social identity theory. This will be done through the example of popular books on evolutionary psychology as well as a supplementary closer examination of what one particular and central author in this debate,

E.O. Wilson, wrote about reductionism in his popular and semi-popular works spanning his career. The final section will then widen the results into a discussion about the use of qualitative evidence in philosophy.

2 Theoretical Background: Social Identity

I will be basing my approach on theories of social identities as developed by Tajfel (1978, 1981) and presented in Hogg and Abrams (1988). Research on scientists' discourses is within Science and Technology Studies frequently theoretically motivated by Gieryn's (1999) concept of "boundary work", which analyses scientists' talk through its functions of delineating social and rhetorical boundaries that mark groups of scientists apart from others. While it has become a paradigmatic theoretical approach within my discipline, I feel however that an approach centred on social psychological concepts of identity are more theoretically developed to provide insights into scientists' philosophical discourses (see Riesch 2010b for a more extensive comparison of boundary work and social identity).

Social identity theory aims to explain intergroup behaviour, relating to prejudice and discrimination, but also the building up of a positive self-concept through the shared representation of what constitutes group membership. Members of a social group enhance their own self esteem by categorizing themselves as conforming to group norms and values. When people categories themselves and others into distinct groups, they tend to overestimate the attributes they have in common with other ingroup members, while underestimating features they have in common with the outgroup.

These perceptions are built up so as to favour the group as opposed to outsiders. Individual group members aim to achieve prestige and status within the group by applying to themselves these desirable group membership criteria, and build up a stereotype of the outgroup(s) by accentuating their perceived negative features. Social groups hold common beliefs, norms, and values that define membership and that individual group members must hold (or at least appear to hold), in order to identify with, and belong to that group (Bar-Tal 1998). The group thus becomes part of the individual.

I argue in this article that especially controversial philosophical concepts that have been caught in heavily debated issues, like reductionism in the Nature/Nurture debates, can easily find themselves being used as membership markers to establish the scientists' philosophical credibility with their peers.

In focusing this essay on the issues of reductionism and the Nature/Nurture controversy, I do not intend to make any judgement or philosophical analysis on the controversy itself, or any of the actors that will be discussed. For my purposes the most important aspects of the debate are that the actors can be clearly divided into two rival groups, and that the debate has more than most scientific controversies been played out through popular science books.

Also I do not intend to judge any of the authors on their usage of the concept of reductionism. As will be shown in the introduction to reductionism below, the definitions of reductionism offered by philosophers have varied greatly, so that no scientist's use of the concept can simply be labelled as philosophically naïve. Most investigations of reductionism and the Nature/Nurture debate like that of Ruse (1989) or even Segerstråle (2000, pp. 284–291) focus on establishing whether what the scientists are doing and or saying is compatible with either their own or otherwise some pre-defined notions of reductionism. This article certainly does not try to do the philosopher's job and 'rescue the empirical scientists from their own philosophical commentaries' (Ruse 1989, p. 58); my focus instead is on establishing the categories and meanings of reductionism that scientists talk about and hopefully offer some insights as to *why* reductionism is talked about as it is. The insights from that analysis will however I hope contribute to the philosophical debate through showing that any deeper exploration of what scientists think of philosophical topics must take into account sociological factors that influence their interpretations; these issues will be discussed in the concluding section: It cannot be simply said that scientists are confused about what the term means, or at least that will not by itself be a very interesting observation. However trying to understand where possible confusions arise will give us an appreciation of the uses of philosophy and how the gaps between philosophers and scientists can be bridged by taking into account various meanings and connotations philosophical terms can acquire when they travel to new interpretive communities. As a qualitative study that pays close attention to what only relatively few scientists have written about reductionism, this paper will not claim to be the final word on the topic and the theory interpretation is to be seen as suggestive—I have obviously no privileged view into what goes on in the scientists' heads. What this paper will do is to suggest a social identity interpretation around an arresting phenomenon that has been noted before by Ruse and Segerstråle, i.e. that self descriptions of being for or against reductionism has followed the delineation between the pro-nature and pro-nurture camps.

This study adopts a discourse analysis approach (Potter and Wetherell 1987) to understand the persuasive and rhetorical dimensions of popular science writing, by questioning the work performed by the philosophy. That philosophical concepts like reductionism have a discursive function, as I suggest in this paper, that of an identity marker, does not however mean that this is their only function. The authors in this study will have thought deeply about the philosophical value of reductionism and where it fits in with their wider philosophical convictions, so they certainly fulfil a more traditionally recognised philosophical function as well. Nevertheless, a qualitative discourse analysis can ask additional questions about the uses of reductionism in the books, such as why is reductionism so much more prominent than other philosophical topics in this particular range of authors, how do they make sense of reductionism, and how does the social and rhetorical function of reductionism outlined in this chapter colour the authors' understanding of it?

3 Reductionism and Popular Science

3.1 Reductionism

Reductionism is frequently recognized to be a very contentious and ill-defined term (see for example Andersen 2001; Ruse 1994; and Dupré 1983). A reductive explanation tries to explain 'higher level' phenomena or theories (such as those of biology) solely by reference to 'lower level' phenomena or theories (such as those of physics). This is one possible reason why some scientists are uncomfortable with reductionism, as it seemingly renders their discipline to be merely a branch of physics: Reductionism is very often seen to be threatening some sciences by reducing them to other sciences. In this context a frequent complaint about reductionism is that it reduces a science (or theory) to *nothing but* another science or theory. In this type of complaint an accusation is often made that reductionism disregards the complexity of the real world. The word 'reductionist' is therefore often used synonymously with 'simplistic'. The reductionist-as-simplistic usage of the term is very prominent especially in the social sciences. On the other side, reductionism is seen to be about the explanation of those complexities using simple or more fundamental premises, and therefore reductionism can be seen as a manifestation of Occam's razor, the principle that we should try to look for the simplest explanation (Ruse 1989, p. 58).

Although reductionism is a concept over which there has been a huge amount of confusion, there are nonetheless real philosophical issues in the different versions of the concept, which I will be categorize below. Philosophical introductions to reductionism often start with Nagel's (1961) model (for example Curd and Cover 1998). For Nagel a reduction is an explanation of a theory by showing that it can be logically derived from another theory. Before Nagel, reductionism has been a concept that featured heavily in logical positivism, but it carried slightly different meanings. One of these for example, the one criticized by Quine, was 'the belief that each meaningful statement is equivalent to some logical construct upon terms which refer to immediate experience' (Quine 1980 [1953], p. 20).

What exactly reductionism is depends on what is to be reduced: in Nagel's case it is theories that are reduced to other theories, in Quine's sense it is theoretical statements that are reduced to observation statements. These are not the only options, and it is debatable if, when one of the popular science authors discussed below talks about reducing facts, it is equivalent to other people talking about reducing phenomena. In general though, I will distinguish roughly between reductionisms that involve theoretical statements, such as models, theories, hypotheses or even whole disciplines, and those involving singular statements of fact or observation, such as facts, phenomena, events. In some definitions of reductionism, theories and facts can be interchanged, but this is not always unproblematic. Thus we can say that both a theory and a fact can be explained by reference to another theory or fact, but the way this should work is not the same. Taking again Nagel's reductionism, we cannot logically derive anything from a singular statement, while reductionism in the traditional logical positivist sense only makes sense when a theory is reduced to singular facts, but not other theories.

Secondly it is not always clear what "reduce" should actually mean. For Nagel, it is quite clearly meant to be a type of explanation: We explain a fact or theory by showing that it follows logically from another theory. Other people can mean it to be more of an ontological statement, that is, a statement about the structure of the world: we are not concerned with whether the scientist actually manages to explain A by B, what matters is that it is theoretically possible. These are of course also linked, because a belief that the world is structured in a reductive way would mean that a successful explanation should reflect that fact.

Thirdly and probably most importantly when we interpret allegiance to reductionism as an identity marker, for each way in which we can define reductionism, there can be two kinds of people who actually call themselves reductionists: We can either claim that science always strives for reductionism, or that it is merely desirable (or that reality is only ever structured in a reductionist way, or just mostly). This particular point is where most of the misunderstandings arise from. People who argue for reductionism generally argue that reductionism is merely desirable, while people who argue against reductionism often claim that reductionism requires science always to be reductionist. For convenience I will call these two options weak and strong reductionism below. The opposite of reductionism is traditionally said to be holism, although I will be trying to limit my use of this concept, as its meaning is as much debatable as that of reductionism, if not more. While most anti-reductionists claim to be holists, and vice versa, there are popular science authors who argue against both (for example Deutsch 1997, p. 21).

3.2 Popular Science and the Nature/Nurture Controversies

There are two reasons for my emphasis on popular science books. Firstly, a popular science book author can express views about science and scientific method unconstrained by the institutional requirements of their technical writing. Secondly, they need to explain aspects of science to the public that they might not feel is necessary to be included in the professional literature, but that the public ought to know. This makes many authors place an emphasis on philosophical topics on scientific method that they would otherwise think are not worth mentioning (see Turney 2001). For example, if an author feels that every scientist is a reductionist, the popular science forum is possibly the only forum where they would feel the need to explain reductionism, because in technical literature they would assume everyone to agree. Popular science offers the author an opportunity to build up an identity as a writer and scientist which is then publicly available and disseminated, to other scientists as well as the public. While popular science provides a relatively unrestrained forum for a scientist to explain their science, it is publicly highly visible and therefore it has frequently been used for the purposes of boundary work (Gieryn 1999) and of fighting out conceptual battles with other scientists (see

Gregory 2003; Fahnestock 1998; Mellor 2003; and especially Cassidy 2006, on the authors considered in this paper, and Nieman (2000), on the use of philosophy for boundary work in popular science).[1]

This paper is the result of pursuing a side-track on a larger study on philosophy in popular science books (Riesch 2008), where I have taken a sample of 30 popular science books written by scientists and shortlisted for the Aventis/Royal Society prize between 1998 and 2004 (Royal Society 2008). These have been analysed closely for their treatment of philosophical topics (see also Riesch 2010a, b, 2012). One early result that stood out prominently was that almost every author (5 of the 30 in total) who was in some way involved in the Nature/Nurture or evolutionary psychology debates mentioned reductionism (the exception being Dawkins, whose book in the sample did not mention reductionism, but who has discussed the topic in earlier books, see below), whereas only 3 of the remaining books in the sample did (these were mainly books covering physics). Furthermore, everyone of the 5 authors sampled who were involved within the debate identified as a reductionist of some form. To analyse this more closely I have decided to look more widely into the debate by looking at other popular science books on the topic. Since the authors in the original sample were either neutral or firmly on the "pro-nature" side, I have looked at how the pro-nurture side have represented reductionism, taking an informal convenience sample of 7 popular books written by the main authors from the other side (Lewontin, Rose and Gould), as identified by other analysts of the debate such as Segerstråle (2000) and Cassidy (2005), taking over also their dividing line between pro-nature and pro-nurture affiliations). Lastly, to look at how the authors' concept of reductionism has developed over time, since the original sample was only a snapshot from the turn of the millennium, I have gone through earlier books written by the protagonists as well. This turned up an interesting shift in the way one particular central author, E.O. Wilson, has written about reductionism. In the analysis Sect. 1 will present in turn the books from the original sample supplemented with some insights about how the same authors have written about reductionism earlier, then I will present how pro-nurture authors have written about it, and finally I will follow the development of reductionism in the books by Wilson.

The sociobiology and evolutionary psychology disputes arise out of a long-standing debate about how much of human behaviour is caused by our nature (or our genes), and how much is caused by our upbringing. One particular semi-popular book which attracted a lot of attention was E.O. Wilson's *Sociobiology: the new synthesis* (Wilson 1975), which has been severely criticized by, among many others, Wilson's Harvard colleagues Richard Lewontin and Stephen J. Gould. Wilson waded into this area in the background of a number of controversies

---

[1]Gieryn (1999) analyses how scientists construct rhetorical boundaries around their areas of professional expertise and thus exclude outsiders form authoritatively commenting and moving into their area. Although Gieryn originally applied his concept to scientists' demarcation between science and non-science, he also argues that it can be applied in disputes within science. Gieryn's boundary approach and the social identity approach that I am relying on in this paper share a certain amount of similarities (Lamont and Molnar 2002; Riesch 2010b).

surrounding the issues of IQ research and eugenics at the time, and so the terms in which these controversies were debated were also very quickly applied to sociobiology. Because sociobiology aimed at explaining the social behaviour of humans through evolutionary mechanisms, it was accused of being an excuse for eugenicists and racists to claim their views as scientifically based. In this earlier debate, reductionism was already one of the contentious terms, as the IQ researchers were accused of reducing humans to nothing but their genes. At about the same time, Richard Dawkins published his book *The Selfish Gene* (Dawkins 2006 [1976]) which also found itself embroiled in the controversy.

In the 1990s a second wave of writers emerged who distanced themselves somewhat from sociobiology itself, but still see their work as part of the tradition started by Wilson and Dawkins, calling their approach 'evolutionary psychology'. These include authors like Steve Pinker and Matt Ridley who together with Wilson himself also featured in my sample, and who have been debating the same opponents, as in the earlier round. Segerstråle (2000) gives an extensive overview of the development of the debates and their origin. Cassidy (2005, 2006) gives an interesting analysis of the later, evolutionary psychology stage of the debate from a science communication point of view.

Settling on a name and a quick account of the dispute itself is somewhat difficult because the fields that concern themselves with the evolutionary study of human behaviour have undergone several name-changes and changes in emphasis. It is probably precisely because this range of subjects has been so controversial that any new developments are dressed up as new disciplines to distance it from its predecessors (see Segerstråle 2000, p. 317). In this paper I have made the decision to refer to the debate as Nature/Nurture. Even though in many ways these terms simplify the dispute itself, I have found talking about being 'pro or anti sociobiology or evolutionary psychology' too clumsy a phrase. Referring to Nature/Nurture may actually have the advantage of putting the debate into the historical context in which reductionism has been so contentious, and reminds us that the controversy did not simply start with the publication of *Sociobiology*.

The Nature/Nurture dispute is also unusual for a scientific dispute because it has also been often fought out within popular science media rather than just the academic circuit. Many of the classic works cited in the debate, such as Dawkins (2003 [1976]) and Wilson (1975) are popular or semi-popular books themselves and therefore very much accessible to the interested layperson. Also the way ideas are discussed in these books is sometimes slightly different to 'regular' popular science. Rather than being an authoritative account of what we know, as popular science can often appear, the books in this debate are often trying to persuade and argue for a particular viewpoint and appear to be often directed at each other as much as at the public. That emphasis on persuasion which is removed from the stylistic constraints of the technical literature may also be a reason why philosophical topics such as reductionism feature more prominently here than in other popular science subjects. These exchanges have of course not exclusively been conducted in the popular science sphere, but still very visibly so (Cassidy 2005, 2006), and certainly much more than is usual in science.

In all of these disputes, the authors can very visibly be divided in two groups, even if only through their opposition to or support of Wilson's earlier work. On one side are popular scientists like Wilson and Dawkins who are joined by a group of popular science writers on evolutionary psychology. On the other side stand Wilson's and Dawkins' principal opponents such as Richard Lewontin, Stephen J. Gould and Steven Rose.

The camps are also visibly divided by their support of, and opposition to, reductionism, as many observers and even some protagonists have pointed out (Segerstråle 2000, Chap. 14; Ruse 1989; Pinker 2002). In this paper I do not intend to add another study that points to this result, although I believe a reading of my sample corroborates it. Instead as mentioned earlier I will below look at the underlying philosophical and sociological message behind the various representations of reductionism, and thereby go beyond just identifying who calls himself a reductionist and who does not.

In the following Sect. 1 will provide small excerpts from the books where the authors have written about reductionism. I have tried to select passages where the authors give a definition or at least a description of what reductionism means. Although I will try to provide the context in which the passage falls where relevant, most often reductionism was mentioned as an aside that had not much particular relevance to the author's wider discussion.

4 Reductionism in the Popular Science Books

4.1 Reductionism in 'Pro-nature' and Neutral Books

The zoologist Ridley (2003) considers himself a reductionist. At the same time, he argues that this position is frequently criticized. In a book that was generally supposedly intended as a peace offering in the debate by arguing how nature works via nurture (as in the title of the book), his argument for reductionism uses a rather sarcastic rhetoric:

> Even to ask such a question reveals me to be a reductionist, and reductionists are BAD THINGS. We are supposed to glory in the holistic experience, and not try to take it apart. (Ridley 2003, p. 163, original emphasis)

This passage shows that Ridley expects the reader to have at least some familiarity with the term already. In fact, it is not until much later in the book that he comes closer to an explanation of what reductionism is, or rather an explanation of what definitely is *not* reductionism. In discussing the sociology of Durkheim, he remarks that

> [Durkheim said:] 'The determining cause of a social fact should be sought among the social facts preceding it and not among the states of individual consciousness.' In other words, he rejected all reductionism. (Ridley 2003, p. 247)

This implies that reductionism means that we should *always* look at the states of individual consciousness to have caused a social fact (Because Durkheim rejected *all* reductionism: If some version of reductionism allowed us to seek another type of cause when appropriate, Durkheim could have accepted it). This seems to correspond to what I called 'strong reductionism' above. Also, Ridley seems to be talking about a reduction being about facts and explanations (as opposed to being about theories and being a statement about the structure of the world). This very strict interpretation is probably not completely fair to Ridley, because this definition of reductionism defines it only by what it is not. In other places he seems to suggest that reductionism is not always appropriate, for example in his book *Genome* he qualifies earlier chapters by admitting that he has 'fallen into the habit of reductionism' (Ridley 1999, p. 148).

Wilson, while also considering himself a reductionist, at least in his book *Consilience* (Wilson 1998), talks differently both about the definition of reductionism and its role in science in an extended passage designed to explain scientific method in general. 'The cutting edge of science is reductionism, the breaking apart of nature into its natural constituents' (Wilson 1998, p. 58). This definition of reductionism represents a subtly different form than the one (apparently) advocated by Matt Ridley. It is enough to reduce as far as naturally possible, but gives us no injunction to look further than that: it is perfectly possible even for a social fact not to have any more natural constituents, and this is therefore a version of weak reductionism, both in the methodological and the ontological sense.

Wilson's initial introduction to reductionism makes his subsequent explanation of how reductionism works somewhat perplexing, because he seemingly equates reductionism with what he perceives as good scientific practice. In one lengthy passage, he gives us an outline of how he thinks reductionism works, 'as it might appear in a user's manual' (p. 58). He then gives an explanation of what seems to be his idea of scientific method which includes some ideas of creativity and objectivity, but does not even actually mention anything reducing to something else. Even more confusingly, on the next page Wilson proceeds by explaining a viewpoint that he confesses to agreeing with, which is a strong version of reductionism because it proposes that *every* law can be reduced. This viewpoint, which he calls 'total consilience',

> holds that nature is organized by simple universal laws of physics to which all other laws and principles can eventually be reduced. This transcendental world view is the light and way for many scientific materialists (I admit to being among them), but it could be wrong. At the least, it is surely an oversimplification. At each level of organization [...] phenomena exist that require new laws and principles, which still cannot be predicted from those at more general levels. (Wilson 1998: 59)

Wilson's attitude towards reductionism therefore seems more complex than he suggests at first. While a weak form of reductionism is so uncontroversial that it is almost the same as scientific method itself, the stronger form of reductionism, which he subscribes to, is recognized as controversial. But actually, even though he admits that he subscribes to the strong version, he immediately qualifies this by saying it is an oversimplification.

Ceccarelli (2001, p. 142) has also remarked on Wilson's equation of reductionism with scientific method. She shows how Wilson keeps changing between strong and weak reductionism throughout his book. She calls this rhetorical strategy 'polysemous textual construction', which she identifies as 'a passage that can be read (that is interpreted) in two or more ways' (p. 5). In this way the author can appeal to different audiences, though in the case of Wilson, Ceccarelli argues that the strategy was unsuccessful and that Wilson was 'uniformly' interpreted by his readers as holding a strong reductionist position (p. 139).

Wilson's concept of *consilience* has often been interpreted as being a version of reductionism itself, though Wilson introduces it as an alternative word for 'coherence' between the sciences (Wilson 1998, p. 6). Also, Wilson's earlier work, the notorious *Sociobiology* (Wilson 1975) itself, has been often seen as a reductionist account due to its aim to connect biology to the social sciences in a similar way to how it is done in *Consilience* (see also Lyne and Howe 1990). As I will show below, while Wilson's motivations might have been the same in the two books, in *Sociobiology* the concept of reductionism not explicitly identified with consilience or coherence between the sciences. On the contrary, in the earlier book Wilson appears to argue *against* reductionism.

Steve Pinker also labels himself as a reductionist although, like Wilson, he argues that there are two forms of reductionism: 'Reductionism, like cholesterol, comes in good and bad forms' (Pinker 2002, p. 69). Of the bad reductionism, Pinker argues that it is not in fact a straw-man, because some scientists have actually held this point of view. Bad reductionism 'consists of trying to explain a phenomenon in terms of its smallest or simplest constituents' (p. 70). That Pinker thinks this is such a misrepresentation of reductionism that it could be misperceived as a straw-man is surprising, because, on the face of it, it is equivalent to Wilson's weaker (and supposedly uncontroversial) reductionism. Pinker, of course, sees himself as a good reductionist: Good reductionism '[...] consists not of replacing one field of knowledge with another but of connecting or unifying them' (p. 70). In the context of his explanation, this seems to mean that Pinker merely requires the different sciences to be complementary and not contradict themselves. This version of reductionism is so watered down that it is hard to imagine how any serious person can disagree with it. Pinker thus almost manages to build up an *anti*-straw-man: this explains Pinker's incredulity that anyone can fail to be a reductionist. But, in fact, we can easily hold an opinion which is similar to Pinker's (i.e. one of the unity of the sciences) and be resolutely anti-reductionist, as long as under reductionism we understand something like Wilson's strong or even weak versions of reductionism.

A similar view of science to the one advocated by Pinker is found in Mayr (1997), in a generally very philosophically orientated book, although Mayr actually arrives at an anti-reductionist conclusion. Mayr stayed more or less neutral in the sociobiology controversies (see Shermer and Sulloway 2004), but explicitly argues against reductionism. Elsewhere (see Ruse 1996, p. 445) he explains that it was his desire to refute reductionism that drove him to write philosophical works in the first place. (Ruse also relates that this desire came from Mayr's opposition to molecular biology; see the discussion on the early Wilson below.) Arguing for the unity of science, he nonetheless cautions against reductionism:

> [A]n advocate of the autonomy of biology might argue in the following way: Many attributes of living organisms that interest biologists cannot be reduced to physicochemical laws, and, moreover, many aspects of the physical world studied by physicists are not relevant to the study of life (or to any other science outside of physics). [...] A unity of science cannot be achieved until it is accepted that science contains a number of separate provinces, one of which is physics, another of which is biology. It would be futile to try to 'reduce' biology, one provincial science, to physics, another provincial science, or vice versa. (Mayr 1997, p. 32)

Mayr's opinion of how science actually works or should work is almost the same as Pinker's good reductionism. Moreover, it is even compatible with Wilson's weak reductionism because Wilson never disputed that some things are not reducible, even though Mayr puts more emphasis on this argument, while Wilson plays it down.

4.2 Reductionism in 'Pro-nurture' Books

Popular science authors who have been writing on the other side of the dispute almost invariably argue against reductionism. Among the most prolific popular science authors who argued against sociobiology is Stephen J. Gould, who also consistently argues against reductionism, devoting a whole chapter to arguing against it (and against Wilson's concept of consilience) in one book (Gould 2003, pp. 198–260). In an earlier book, when discussing Dawkins, he remarks:

> I think, in short, that the fascination generated by Dawkins's theory arises from some bad habits of Western scientific thought – from attitudes [...] we call atomism, reductionism, and determinism. The idea that wholes should be understood by decomposition into 'basic' units; *that properties of microscopic units can generate and explain the behavior of macroscopic results* [emphasis added]; that all objects have definite, predictable, determined causes. (Gould 1980, p. 77)

Here, Gould associates Dawkins' theory not merely with reductionism, but also with atomism and determinism as 'bad habits of Western thought'. While his description of atomism can be seen as a version of ontological reductionism (as described above), determinism and reductionism are independent philosophical viewpoints which are very often associated together by writers opposed to both of them (materialism and mechanism can be added to this list as well). It could be argued that this association is predisposing people who wish to argue against determinism, against reductionism, so that the author ends up believing reductionism to be whichever version accords closest to his/her anti-determinist views. Reductionism here is taken to mean the belief that *at least some* microscopic units can 'generate and explain the behavior of macroscopic results'. In other words, Gould seemingly believes this can never be the case.

However, in other works a slightly more relaxed anti-reductionism is favoured:

> The depth [of determinism] records the link of biological determinism to some of the oldest issues and errors of our philosophical traditions – including *reductionism*, or the desire to explain partly random, largescale, and irreducibly complex phenomena by deterministic behavior of smallest constituent parts (physical objects by atoms in motion, mental functioning by inherited amount of central stuff). (Gould 1992, p. 27 original emphasis)

Here there is no suggestion that it is not at least sometimes permissible to explain reductively, because we're not talking of *any* macroscopic results, but only irreducibly complex ones. The implications are that Gould is in fact allowing the occasional reductive explanation, where appropriate. Note again Gould's association of determinism and reductionism.

Other anti-reductionists however are less strict with what they allow as permissible science. Steven Rose shows his anti-reductionist credentials by remarking that: '…I have spent a considerable portion of my theoretical energies over the years criticizing reductionism' (Rose 1992, p. 210). As an explanation of what reductionism means he remarks that it possesses an 'insistence that in "the last analysis" the world can be explained in terms of atomic/quantum properties and a few universal assumptions' (p. 74). This would make Rose's version of reductionism an even stronger one than Pinker's bad reductionism, i.e. that *everything* ultimately is explainable by fundamental physics—and therefore something that Pinker might consider an unfair straw-man representation of reductionism.

Situating this description of reductionism with a discussion of Descartes' materialist philosophy, Rose consistently associates reductionism with mechanism: '…I could only be amazed at how deeply my thinking had become trapped into a mechanistically reductionist straightjacket' (p. 287). Interestingly, Rose distinguishes between methodological and philosophical reductionism, with only the latter being the subject of his criticism. Methodological reductionism is to 'try to stabilize the world that one is studying by manipulating one constant at a time, holding everything else as constant as possible' (p. 210). I do not, however, see any connection between Rose's 'methodological reductionism' and any of the forms of reductionism I have been discussing so far. Nor does it seem to prevent Rose from consistently describing himself as being against reductionism. It is interesting then that Rose, just like Wilson and Pinker, identifies a form of reductionism he agrees with—but unlike them does not think that would make him a reductionist.

Finally, Richard Lewontin also argues against reductionism. Here Lewontin is talking about an ontological position rather than explanation or a methodological prescription of how to do science. Reductionism as a methodology is understood to be of the very strong kind, i.e. that every explanation has to be reductive.

> Now it is believed that the whole is understood *only* by taking it into pieces, that the individual bits and pieces, the atoms, molecules, cells, and genes, are the causes of the properties of the whole objects and must be separately studied if we are to understand complex nature. (Lewontin 1993, p. 12, original emphasis)

Like Rose associates reductionism with materialism, Lewontin associates reductionism with mechanism, he writes for example about '[t]he difficulties of the reductionist mechanical view of biology...' (Lewontin 2000: 45). Ironically, it was Lewontin (writing together with Richard Levins) who, 15 years earlier, expressed impatience with the constant confusion of reductionism and materialism (Levins and Lewontin 1985, p. 133).

4.3 The Development of Reductionism in the Works of E.O. Wilson

Even though Wilson argues for reductionism in *Consilience*, he has actually also been described as 'holistically orientated' (Segerstråle 2000, p. 290). In fact, Wilson has undergone quite a change in his talk about reductionism, and this otherwise puzzling transformation can possibly be explained by the different groups that Wilson has been battling and belonging to over the years. Before the publication of *Sociobiology*, for example, there has been a real fear among zoologists that the evidently 'reductionist' discipline of molecular biology is going to replace mainstream biology (see his autobiography, Wilson 1994, Chap. 12). So, reductionism does not always seem to have been something positive for Wilson. This is reflected in an early book on insect societies where it appears only in association theorists he disagreed with:

> Albrecht Bethe, an extreme reductionist, believed that ants are 'reflex machines' [...]. Theodore C. Schneirla [...] took a position as close to the opposite as was possible. [...]. His intent, I believe, was also reductionist [...]. It is now very clear that neither of these opposing simplistic schemes accurately identified the innate and experiential elements of behavior. (Wilson 1971, p. 221)

Note Wilson's association of reductionism with simplistic explanation.

Later, in the beginning of his famous book *Sociobiology*, he talks about a 'new holism', how it stands in direct contrast to reductionism, and how his study is meant to be holistic. I think there are quite a few things that stand out in this quote, so I have numbered different passages for ease of analysis.

> [1] The recognition and study of emergent properties is holism, once a burning subject for philosophical discussions [...], [2] but later, in the 1940's and 1950's, temporarily eclipsed by the triumphant reductionism of molecular biology. [3] The new holism is more quantitative in nature, supplanting the unaided intuition of the old, it does not stop at philosophical retrospection but states assumptions explicitly and extends them in mathematical models that can be used to test their validity. [4] In the sections to follow, we will examine several properties that are emergent and hence deserving of a special language and treatment. (Wilson 1975, p. 7)

First of all, passages one and four together show that Wilson means his work to be holistic, because his book will deal with emergent properties, the study of which he says is holism. In passage two he identifies reductionism as a conceptual rival to holism, but one which has had its day. Most interesting I find his description of 'new' holism (passage three) which, just like his later description of reductionism, actually seems to be his idea of good scientific method. This description of holism as studying emergent phenomena is, remarkably, very much compatible with the position he will later identify with 'weak' reductionism and (but this depends on his treatment of emergent phenomena), even possibly 'strong' reductionism. Finally, (in passage two) he also reveals that his criticism of reductionism is directed at molecular biology.

Only three years after the publication of this book, with the row it engendered now in full swing, he starts talking differently about reductionism. Though he still thinks there is more to science than 'raw' reductionism, he no longer places it in opposition to his own holism, but sees it rather as complementary. He starts by approving of Mach's reductionist philosophy, and then adding on to it:

> The heart of the scientific method is the reduction of perceived phenomena to fundamental, testable, principles. [...] Although Mach's perception has an undeniable charm, raw reduction is only half of the scientific process. The remainder consists of the reconstruction of complexity by an expanding synthesis under the control of laws newly demonstrated by analysis. (Wilson 1978, p. 11)

He ends this passage by lamenting that 'Reduction is the traditional instrument of scientific analysis, but it is feared and resented.' (p. 13). Wilson here seems to have philosophically been fighting two enemies. First, against molecular biology, which is traditionally described as reductionist (because it explains biological phenomena by what happens at the molecular level), he identifies as a holist. When, in the sociobiology debates, he gets accused of reductionism, he becomes a reductionist. Note that I am not accusing Wilson of inconsistency here, as he is perfectly entitled to change his opinion. In fact, Wilson has noted (1994, Chap. 12) that over the years he came to see many of the molecular biologists' points, and reductionism seems to be one of them.

But now Wilson talks as if he had *always* been a reductionist. Writing about the 1960s in his autobiography, he states that 'I believed deeply in the power of reductionism, followed by a reconstitution of detail by synthesis' (Wilson 1994, p. 255). This suggests that Wilson must think that in the early 70s he was only wrong about his *definitions* of holism and reductionism, rather than about the content of his opinion on how science works.

5 Analysis

5.1 Philosophical Opinions and Philosophical Identities

Although there really do seem to be some fundamental differences about both how science should work and about how the world is structured, these differences do not come out very clearly in the author's discussion about reductionism, particularly because these divisions do not run along the lines of who calls himself a reductionist and who does not. If we take reductionism to mean that strong reductionism (i.e. that *all* explanations have to be of a reductive nature), as most of the anti-reductionists argue, then only Ridley is possibly a reductionist, while Wilson (in *Consilience*) appears ambivalent. On the other extreme, Pinker has ideas about science that are seemingly less reductionist that those of the anti-reductionists I've quoted. It is interesting then to notice that at least by the time Wilson writes *Consilience*, he identifies so strongly with being a reductionist, even if his actual definition of reductionism is seemingly at odds with some of the other reductionist authors writing on his side of the dispute. Another curiosity is the similarity between some declared reductionist and anti-reductionist positions, like those shown above between Mayr and Pinker.

The way these authors see and describe themselves as being either pro or anti-reductionism despite the way they actually define reductionism is part of how they see themselves philosophically and how they would like others to see them. The authors are writing popular science which is aimed not just at each other, both pro and contra sociobiology, but also at the general public. This possibly gives extra incentive for the author's positioning of themselves as part of a philosophical group which they find socially desirable to belong to. To construct their social identity and belong to the group, they need to know which values and beliefs they must acquire and convince others that they have them (Bar-Tal 1998). A commitment or opposition to reductionism could be such a belief, following the rhetoric for the opposing camps in the sociobiology dispute. Then, rather than being a mere philosophical or methodological commitment, reductionism or opposition to it marks the authors' membership of what they perceive is a socially desirable group.

In social identity theory the individual self-esteem of members within a group is satisfied by 'maximizing the difference' (Hogg and Abrams 1988, p. 23) with outsiders. Likewise, differences within the group can be downplayed. Seeing how strongly the opposition to sociobiology identified Wilson and those defending him with reductionism, it is very possible that he started identifying with being exactly that. Thus it was possible for Wilson the holist to become Wilson the reductionist, without him actually having to change his opinion.

This development, I think, adequately describes how Wilson could have come to see himself as a reductionist, but still leaves open why Wilson did not revise his actual opinion on the matter, which, as I hope to have demonstrated, is not so very much different to other people who do not think they are reductionists, and even his older, holistic self. In other words, although Wilson's idea about what reductionism is could well have changed through the mechanisms of social representation and social identity, his underlying philosophy seems to have stayed fairly constant.

5.2 Conflicts Between Philosophical Identity and Philosophical Opinion

As shown above, many people, especially the critics of reductionism, associate reductionism with determinism, materialism and atomism. This fact is not only noticed by Levins and Lewontin, as mentioned above, but also Richard Dawkins, who observed, when called a genetic determinist, that determinism 'is one of those words like sin and reductionism: if you use it at all you are against it.' (Dawkins 2003, p. 197). In an endnote to the newer editions of 'The selfish gene' (Dawkins 2006 [1976], p. 331), he also displays his impatience with arguments by his opponents against reductionism, and that the only good arguments they offer are against determinism rather than reductionism. However, he sees himself as a reductionist, but not a determinist.

Philosophical viewpoints very often come together as packages, even if they assert things independently. A commitment to classical logical positivism for example would also entail a commitment towards reductionism (in Quine's sense), anti-realism and other things. Segerstråle (2000, pp. 284–291) examines how the critics of sociobiology typically charged the 'Nature' side with being reductionist and how this usually came from an ideological worldview in which reductionism is a sign (among others) of bad science. Thus, unless they indulge in a specific philosophical criticism, followers of a particular philosophical school might have to accept (reject) philosophical views they would otherwise (not) agree with. This is not merely a case of swallowing a bitter pill. Rather it represents a distortion of what the scientist thinks a philosophical viewpoint, which he/she might not otherwise agree with, is actually about. For example, consider people who sees themselves unshakably as a Marxist, and then learn that all Marxists are materialists (suppose they do not really know what materialism is, but it is not something they would agree with if they did), then there will be a tension between their identity as a Marxist and their actual opinion. One possible way of solving this tension is to believe in a watered-down version of materialism that they can support, and believe that all other accounts are straw-men.

In a similar vein, I think there is a tension evident in the popular science writing on the Nature/Nurture debate. As it is desirable for the authors who chose a side in the debate to define themselves philosophically consistent with the rest of that social group as either reductionist or anti-reductionist, they must either change their opinions to fit with what they believe to be reductionism, or adopt a type of reductionism that accords with their opinion. In the case of such an already highly abstract, intangible and contested concept as reductionism, the latter option is certainly understandable.

In this paper I hope to have shown how the different sides in the dispute take sides with regard to reductionism, even if their actual opinion of the philosophical issues underlying reductionism do not necessarily reflect this division perfectly—there are many different ways of interpreting reductionism and therefore it is easy for a scientist to identify with (anti)reductionism without at least substantially changing philosophical opinion about science. This is also reflected in the way Wilson's talk about reductionism has changed. Though the general messages of *Sociobiology* and *Consilience* may have been similar regarding the unity of science, his actual talk about reductionism is different now than it was then. It could be argued that Wilson was employing a polysemous strategy in *Sociobiology*, as Ceccarelli has argued for *Consilience*, so that he combined both an anti-reductionist talk with an alternating reductionist message, just as he was alternating weak and strong reductionism in *Consilience*. But even then there is a different talk about reductionism, from at least seeming to reject it, to accepting it and being vague about the details.

All this is not to say that reductionism is central to the authors' identities as being part of their side in the debate, nor is it to say that declaring oneself to be an (anti-) reductionist is necessary in order to proclaim which side they are on-this is often stated clearly enough anyway. However, (anti-) reductionism is part of the group of norms, values and philosophical convictions that make up social identity for either side, while not central as such, it is nevertheless strong enough for almost every author involved in the debate to mention it and proclaim an affiliation at some point.

6 Conclusions

A fuller analysis on the debate will uncover many other, more substantial, identity markers for the two sides, however, as an analysis of one particular philosophical concept, this chapter aims to demonstrate how it can have additional social functions beyond its philosophical value.

The way philosophical opinions about science are expressed in popular science can follow social and disciplinary divisions among the scientists as well as reflect a deeper philosophical commitment. At the same time, who calls himself a reductionist and who does not may be more influenced by disciplinary identity than actual philosophical opinion. This is best illustrated when the argument is made public as spectacularly as it is in the Nature/Nurture disputes. Of course it is not only philosophical beliefs that can become identity markers for a scientific group, however, because of their high abstraction, and the general lack of (philosophers') agreement about their precise meaning, they are particularly vulnerable to a re-interpretation that accords well with the scientists' other social boundaries. This interpretation of what is going on with respect to scientists' opinions on reductionism is somewhat re-inforced by an accompanying study I performed consisting of interviews with working scientists (Riesch 2008), none of whom had been involved in these debates, and who almost unanimously found the topic of reductionism rather uninteresting.

This points to a lesson for any philosophical analysis that intends to use qualitative evidence to find out how scientists (or anyone else) thinks about philosophical issues. It is not enough to look at how philosophical topics are talked about, certainly not in a simplistic sense of which scientists consider themselves reductionists or anti-reductionists because this leaves out the complex social and rhetorical positioning that scientists as much as anyone else engages in. Philosophical context gets reinterpreted and made use of in the many other social ways through which philosophical talk is useful for scientists, and keeping this in mind will help us avoid interpreting scientists' philosophies along ways that does not actually accord with their philosophical opinion—it will merely reflect the philosophical position they think they hold, or if we're more cynical, the philosophical position that they feel it is desirable to hold.

A purely social science analysis may well find this insight interesting in its own right, and see it as (yet) another case of scientists constructing rhetorical boundaries around their social groups. However a philosophical analysis may want to do more, for example to use scientists' philosophical ideas to inform the philosophy itself (for example, Bailer-Jones 2003). My study I hope has shown that scientists' philosophical talk cannot be interpreted without grounding it in a sociological theoretical context, as otherwise the nuances in thinking and expression will be lost and philosophically careless remarks which are designed for completely different purposes than philosophical theorising may inadvertently get too much attention. At the end, the use of philosophy for scientists is much greater than merely philosophising.

References

Andersen, H.: The history of reductionism versus holistic approaches to scientific research. Endeavour **25**(4), 153–156 (2001)
Bailer-Jones, D.M.: Scientists' thoughts on scientific models. Perspect. Sci. **10**, 275–301 (2003)
Bar-Tal, D.: Group beliefs as an expression of social identity. In: Worchel, S., Morales, J.F., Páez, D., Deschamps, J. (eds.) Social Identity: International Perspectives, pp. 93–113. Sage, London (1998)
Cassidy, A.: Popular evolutionary psychology in the UK: an unusual case of science in the media? Publ. Underst. Sci. **14**(2), 115–141 (2005)
Cassidy, A.: Evolutionary psychology as public science and boundary work. Publ. Underst. Sci. **15**(2), 175–205 (2006)
Ceccarelli, L.: Shaping Science with Rhetoric: The Cases of Dobzhansky, Schrödinger, and Wilson. University of Chicago Press, Chicago (2001)
Chang, H.: Inventing Temperature. Oxford University Press, Oxford (2004)
Curd, M., Cover, J.A. (eds.): Philosophy of Science: The Central Issues. W. W. Norton, London (1998)
Dawkins, R.: A Devil's Chaplain. Weidenfeld & Nicolson, London (2003)
Dawkins, R.: The Selfish Gene, New edn. Oxford University Press, Oxford (2006)
Deutsch, D.: The Fabric of Reality. Penguin, London (1997)
Dupré, J.: The Disunity of Science. *Mind* xcii, pp. 321–46 (1983)
Fahnestock, J.: Arguing in different forums: the Bering crossover controversy. Sci. Technol. Human Values **14**(1), 26–42 (1998)
Gieryn, T.F.: Cultural Boundaries of Science: Credibility on the Line. University of Chicago Press, Chicago (1999)
Gould, S.J.: The Panda's Thumb. Penguin, London (1980)

Gould, S.J.: The Mismeasure of Man. Penguin, London (1992)
Gould, S.J.: The Hedgehog, the Fox, and the Magister's Pox: Mending the Gap Between Science and the Humanities. Jonathan Cape, London (2003)
Gregory, J.: The popularization and excommunication of Fred Hole's "life-from-space" theory. Publ. Underst. Sci. **12**(1), 25–46 (2003)
Hogg, M.A., Abrams, D.: Social Identifications. Routledge, London (1988)
Lamont, M., Molnár, V.: The study of boundaries in the social sciences. Ann. Rev. Sociol. **28**, 167–195 (2002)
Levins, R., Lewontin, R.: The Dialectical Biologist. Harvard University Press, Cambridge, MA (1985)
Lewontin, R.: The Doctrine of DNA. Penguin, London (1993)
Lewontin, R.: It Ain't Necessarily So: The Dream of the Human Genome and Other Illusions. Granta, London (2000)
Lyne, J., Henry, H.: The Rhetoric of Expertise: E.O. Wilson and Sociobiology. Q. J. Speech **76**(2), 134–151 (1990)
Mayr, E.: This is Biology. Harvard University Press, Cambridge, MA (1997)
Mellor, F.: Between fact and fiction: demarcating science from non-science in popular physics books. Soc. Stud. Sci. **33**(4), 509–538 (2003)
Nagel, E.: The Structure of Science. Routledge, London (1961)
Nieman, A.: The popularization of physics: boundaries of authority and the visual culture of science. Unpublished PhD thesis, University of the West of England (2000)
Pinker, S.: The Blank Slate. Penguin, London (2002)
Potter, J., Wetherell, M.: Discourse and Social Psychology: Beyond Attitudes and Behaviour. Sage, London (1987)
Quine, W.V.O.: From a Logical Point of View, 2nd edn. Harvard University Press, Cambridge, MA (1980)
Ridley, M.: Genome. Fourth Estate, London (1999)
Ridley, M.: Nature via Nurture. Fourth Estate, London (2003)
Riesch, H.: Simple or simplistic? Scientists' views on Occam's razor. THEORIA Int. J. Theo. Hist. Found. Sci. **25**(1), 75–90 (2010a)
Riesch, H.: Theorizing boundary work as representation and identity. J. Theor. Soc. Behav. **40**(4), 452–473 (2010b)
Riesch, H.: On the philosophical talk of scientists. In: Francois, K., Loewe, B., Mueller, T., van Kerkhove, B. (eds.) Foundations of the Formal Sciences VII. College Publications, London (2012)
Riesch, H.: Scientists' views of the Philosophy of Science. Unpublished PhD, University College London (2008)
Rose, S.: The Making of Memory. Bantam, New York (1992)
Royal Society: Royal society prizes for science books. Retrieved March 18, 2009, from http://www.royalsoc.ac.uk/sciencebooks (2008)
Ruse, M.: Sociobiology and reductionism. In: Hoyningen-Huene, P., Wuketits, F. (eds.) Reductionism and Systems Theory in the Life Sciences, pp. 45–83. Kluwer, Dordrecht (1989)
Ruse, M.: Knowledge and human genetics: some epistemological questions. In: Weir, R., Lawrence, S., Fales, E. (eds.) Genes and Human Self-Knowledge. University of Iowa Press, Iowa City (1994)
Ruse, M.: Monad to Man: The Concept of Progress in Evolutionary Biology. Harvard University Press, Cambridge, MA (1996)
Segerstråle, U.: Defenders of the Truth. Oxford University Press, Oxford (2000)
Shermer, M., Sulloway, F.: The grand old man of evolution: an interview with evolutionary biologist Ernst Mayr. Skeptic **8**(1), 76–82 (2004)
Tajfel, H. (ed.): Differentiation Between Social Groups: Studies in the Social Psychology of Intergroup Relations. Academic Press, New York (1978)
Tajfel, H.: Human Groups and Social Categories: Studies in Social Psychology. Cambridge University Press, Cambridge (1981)

Turney, J.: More than story telling—reflecting on popular science. In: Stocklmayer, S., Gore, M., Bryant, C. (eds.) Science Communication in Theory and Practice. Kluwer, Dordrecht (2001)
Wilson, E.O.: The Insect Societies. Belknapp Press of Harvard University Press, Cambridge, MA (1971)
Wilson, E.O.: Sociobiology: The New Synthesis. Belknapp Press of Harvard University Press, Cambridge, MA (1975)
Wilson, E.O.: On Human Nature. Harvard University Press, Cambridge, MA (1978)
Wilson, E.O.: Naturalist. Penguin, London (1994)
Wilson, E.O.: Consilience. Abacus, London (1998)

An Empirical Method for the Study of Exemplar Explanations

Mads Goddiksen

Abstract The most common way of studying explanations in philosophy of science and science education is through case studies. Recently these have been supplemented with studies based on empirical methods. This chapter provides an empirical method for collecting and comparing exemplar explanations across scientific disciplines with the aim exposing possible qualitative differences between them. The method is based on the use of science textbooks as sources of explanations. I discuss a number of possible strategies for identifying explanations in these sources, and specify a set of reliable linguistic indicators that can be used for this purpose. A pilot study is presented to illustrate the method and its limitations.

Keywords Explanation · Exemplars · Text books

1 Introduction

Within philosophy of science there has been considerable interest in scientific explanations for several decades. There is general agreement that constructing and evaluating explanations is a very important part of what practicing scientist do, but there is less agreement on *how* and *why* they do it. Many of the classical studies of scientific explanations aim to answer the question of how scientific explanations differ from non-scientific explanations by providing accounts of the characteristics of scientific explanations regardless of which part of science they originate from (Friedman 1974; Hempel 1965; Salmon 1998). Against this overall project Van Fraassen argued that there are no interesting common features in explanations across all the sciences (Van Fraassen 1980). Others have argued that although explanations are important in all the sciences, the standards for what counts as a *good* explanation

---

M. Goddiksen (✉)
Department of Physics and Astronomy, Centre for Science Studies, Aarhus University, Aarhus, Denmark
e-mail: madsgoddiksen@yahoo.dk

can change as disciplines change over time (McMullin 1993) and it is widely recognized that disciplines coexisting at one period of time have different standards for what counts as a good explanation (Godfrey-Smith 2003, Chap. 13; Woodward 2011; Woody 2003). This opens for more specific studies of explanations from specific disciplines [e.g. mechanistic explanations in the life science (Machamer Darden and Craver 2000)] and comparative studies across disciplines.

Given that explanations play a key role in scientific practice, learning to construct and evaluate explanations should also be an important part of any science education. This has also been recognized within the science education literature (Braaten and Windschitl 2011) and researchers in this field have therefore taken an interest not only in scientific explanations given by scientists to their peers and how these differ from every day explanations but especially in explanations intended for students. One conclusion from this research is that there is a need for more explicit teaching on how to construct high quality explanations (see e.g. Solomon 1995; Peker and Wallace 2011).

Many studies in the science education literature refer to studies from other disciplines such as philosophy, linguistics or discourse analysis (Edgington and Barufaldi 1995; Rowan 1988; Unsworth 2001) as a general framework for the development of a more detailed analysis of explanations from textbooks or teaching situations. It is interesting to note that the discussions about the potential paradigm dependence of the criteria for high quality explanations have been somewhat overlooked in the science education literature. This means there is a risk that explicit teaching in how to construct good explanations will not be sufficiently nuanced.

In this chapter I outline a methodology that can be used to study and compare explanations from different disciplines drawing on both science education and philosophy of science. More specifically, the main aim of this chapter is to formulate an empirical method based on gathering and analyzing the *exemplar explanations* practicing scientists use in education. Following Kuhn (1996), exemplar explanations are explanations presented to coming members of a scientific community both in order to help them understand a given explanandum, but also to display what a good explanation looks like (see also Treagust and Harrison 1999). These explanations thus play an important role in teaching coming members of a scientific community how to construct good explanations. By comparing the exemplar explanations it is possible to identify and describe possible differences in standards for good explanations across disciplines. The method presented in this chapter is therefore developed to provide a descriptive account of exemplar explanations that is sensitive both to differences in practices between different disciplines and to differences in practices between different educational levels. As I have argued elsewhere (Goddiksen 2013), the results of such a comparison will, for instance, be valuable to educators aiming to teach interdisciplinary problem solving. One of the epistemological challenges faced in interdisciplinary problem solving is to navigate the differences in standards for good explanations across different disciplines and knowing what these differences are will *ceteris paribus* make this process easier.

To argue for the value of empirical studies of explanations, and to see how my method differs from previous empirical approaches to the study of explanations,

I start out by discussing Andrea Woody's empirically based account of explanations in chemistry (Sect. 2) and the methodology presented by Zoubeida Dagher for the study of explanations given by science teachers (Sect. 3).

2 Andrea Woody's Account of Explanations in Chemistry

Empirical studies of explanations are rare in the philosophy of science literature. An important exception is the empirically based account of explanation in chemistry that has been developed by Woody (2004a, b) who has examined examples of explanations in a mainstream chemistry textbook.[1] Although the aim of Woody's study of explanations is quite different from mine, it is interesting to discuss her general argument for choosing an empirical method.

2.1 Why Choose an Empirical Method

Woody has chosen an empirical method for two reasons. One is her wish to give a highly descriptive account of explanations in chemistry (Woody 2004a, pp. 17–18). The other reason is that she sees a flaw in the argumentation in earlier case based studies that she wants to avoid.

According to Woody the typical way to analyze explanations in philosophy of science has been through inductive arguments starting from a small set of paradigmatically successful explanations (Woody 2004a, p. 36). The structure of these inductive arguments can roughly be represented by what we might call the classical pattern of argument or just CP (ibid.):

1.1. a is a successful explanation of b
1.2. The basic, or most noteworthy, characteristics of a are $\{j, k, l, \ldots\}$.
1.3. Members of the set $\{j, k, l, \ldots\}$ are (quasi)tokens of corresponding types $\{J, K, L, \ldots\}$
   Infer by generalization
   C_{CP}: The requirements for successful explanation are $\{J, K, L, \ldots\}$

C_{cp} can then be tested by analyzing other paradigmatically successful explanations. Conversely, the easiest way to argue against conclusions drawn from CP is to find a paradigmatically successful explanation c whose characteristics are not (quasi)tokens of $\{J, K, L, \ldots\}$.

---

[1]Woody actually calls her method "quasi-empirical" (2004a, p. 13). This seems reasonable given that her empirical material is limited to just one textbook. However there is nothing semi-empirical about the methodology employed in the study.

Woody's critique of this way of analyzing scientific explanations is that the justification for the choice of paradigmatically successful explanations is deficient. This would not be a problem if it was possible to identify a relatively large set of candidates that everyone (or at least all philosophers of science) would intuitively accept as successful explanations. But unfortunately we do not have such a set. This has led to what Woody sees as a rather pointless debate:

> [Philosophers] quarrel famously over a set of reputed, but still disputed, "counter-examples": the flagpole and the shadow, the ink spill on the carpet, leukemia and radiation exposure, John Jones' recovery from pneumonia. This dispute cannot possibly be settled in this manner. (Woody 2004a, p. 15)

It seems then that claiming that a certain case is a good example of a successful explanation is far from trivial. This means that the choice of cases needs additional justification. According to Woody (2004a, p. 39), the only *theoretical* justification that can be offered for the choice of the examples is an appeal to pre-analytic intuitions about the general nature of (successful) explanations. However, Woody argues that this kind of justification would make the account viciously circular:

> It is precisely the general nature of explanation […] we are attempting to determine via this argument. Thus either we are involved in a vicious form of circular reasoning or we need some independent means of justifying [this premise]. (Woody 2004a, p. 39)

Hence Woody's second argument for choosing an empirical method is to avoid this kind of vicious circular reasoning (hereafter referred to as Woody's circularity objection). Here an empirical method means identifying sources of explanations that practicing scientists deem successful and extract the explanations from these. This way, according to Woody, it is possible to avoid the kind of vicious circular reasoning involved in the earlier studies.

The question is of course (a) what sources to choose, and (b) how to identify the explanations in the sources? Woody's answer to (a) is that science textbooks are highly useful sources because they "articulate the most common explanatory strategies of a discipline, often accompanied by implicit suggestions regarding explanation's role within the discipline" (Woody 2004a, p. 18). But at the same time Woody stresses that one also needs to keep in mind that also textbooks may vary in their perspective on the disciplines, and that their descriptions may vary from actual practice due to their educational aims.

With respect to (b), Woody does not give an explicit answer to how explanations in textbooks can be identified. I agree with Woody that science textbooks are valuable sources when studying scientific explanations empirically, and in this chapter I shall develop a method that identifies explanations from a given discipline with a certain *quality*, namely explanations that are accepted by practicing scientists from this discipline as good explanations to give to students.

Admittedly, if the identification of these explanations is based on pre-analytic intuitions about the nature of explanations with this quality the study would still be vulnerable to Woody's circularity objection. But I shall argue that such assumptions need not be made. As outlined below I suggest is that the focus should rather be on identifying explanations in sources that we have independent reason to believe contain *only* explanations with the desired quality. Thus I avoid the vicious *normative circle* where a normative claim about the nature of *good* explanations to give to students is based on intuitions about what constitutes a *good* explanation to give to students. However the argument is still based on a *descriptive* circle. I will still have to assume something about the nature of explanations in order to identify them in the sources. Thus the description of the explanations in the sources must still rely on an explicit or intuition based pre-analytic description of explanations that enables us to recognize explanations in the sources, so clearly the method developed here involves circular reasoning. But this kind of descriptive circularity, which is present in any empirical investigation, is not vicious (Nersessian 1995).

So if Woody's objection is interpreted as a reference to only the vicious *normative* circularity involved in earlier studies, the method developed here will not be targeted by the objection. This of course will only be the case if satisfactory answers to questions (a) and (b) can be provided along with independent reasons for why the chosen sources contain only the desired kind of explanations.

2.2 Identifying Good Sources

What would be a good source of exemplar explanations given to students by practicing scientists from a given discipline? Since my primary interest here is in explanations that are *widely accepted* within the discipline as being of high quality it seems reasonable to look mainly to written sources that have been through some kind of critical review. These kinds of written sources fall into two general categories: (1) peer reviewed journal articles and other documents that aim to convey novel results to practicing scientists, and (2) textbooks and other documents that aim to convey established knowledge from the discipline to students (among others).

Although an explanation presented in a journal article has been through peer review, it is not necessarily uncontroversial. Some articles do reach such a high status within a discipline that they become widely used as prototypes of what a good scientific article and a good scientific explanation is. If these can be identified they might prove valuable, but the primary source of exemplar explanations are the widely used textbooks from the disciplines under investigation.[2] These sources are written explicitly with teaching in mind and all the explanations in them have been

---

[2]Textbooks are not only particularly suited for studies of explanations to students, they are also more generally good sources of explanations. Indeed, anyone interested in widely accepted explanations should be interested in textbooks, since the explanations found in scientific articles are not necessarily uncontroversial. Furthermore textbooks are useful for a study (like Woody's) that aims to answer *why* explanations are so important in scientific practice, because textbook explanations can provide clues as to why and how explanations are valuable to practitioners since one of the aims of a textbook is to show future practitioners how to use the tools of the discipline (Woody 2004a, p. 18).

carefully selected as suitable explanations to give to students at a given level of education. These explanations are thus constrained both by the educational level of the intended audience and the standards for good explanations in the given discipline (Treagust and Harrison 1999). Arguably, the constraints from the level of the intended audience are most prominent in lower level textbooks, whereas most advanced textbook are primarily constrained by the standards for good explanations in the relevant discipline. In order to gain knowledge about how practicing scientists' explanations to students develop as the students progress through their education, it will therefore be important to make sure that the selection of textbooks includes both introductory and advanced texts.

2.3 How to Identify Explanations in the Sources Selected?

Once a set of sources is identified, the next step is to identify the explanations in them. This is complicated by the fact that although a textbook may contain many explanations of natural phenomena, experimental procedures etc., they do not necessarily contain *just* explanations. For one a textbook may have to devote space to describe various explananda, which may include the description of experimental setups and puzzling data. Furthermore, many explanations are cashed out in terms of entities and activities, some of which may not be familiar to the student (Machamer et al. 2000). Significant portions of a textbook may thus have to be devoted to introducing entities (abstract and concrete) or activities that feature in explanations, but are not necessarily explained themselves to a significant extent. For instance: In order to be able to explain the workings of a Scanning Tunneling Microscope it will be important to introduce the activity "quantum tunneling" as well as entities such as electrons that perform these activities. Ogborn and collaborators thus found that in the classroom, science teachers in secondary school spend much time on "the construction of entities" (Ogborn et al. 1996, Chap. 3) used in explanations. Looking closer at their examples, reveals that this often involves introducing the activities they are involved in as well (p. 39). In addition to entities and activities that are parts of mechanisms, textbooks may also introduce a number of models that can be used as parts of explanations of other things. Some of these introductions will be explanations, but it is not possible to explain everything. Some entities, activities and models are likely to be introduced as black boxes that may or may not be explained elsewhere. When studying textbooks the investigator therefore cannot take for granted that every passage of a textbook in meant to explain and provide understanding. Passages that are not meant to be explanatory need not live up to the standards of good explanations in place within the discipline, and including them in the sample may thus give a distorted picture of what is considered a good explanation to give to students. It is therefore necessary to have some form of criteria that can be used to identify the explanations that are contained in the sources. There seem to be a number of ways in which an investigator may approach this challenge. One is to rely on pre-analytic intuitions about good explanations to give to students. Relying on these would of course make the empirical argument just as viciously circular as the case-based studies that Woody criticizes. There is thus a general worry that simply making an empirically based claim is not in itself sufficient to overcome Woody's circularity objection. If the claim is based on a biased dataset gathered using a spurious method it should not be considered any better than an invalid theoretical argument. In order to overcome Woody's circularity objection it is therefore important that a way of identifying explanations in textbooks is found, that is not dependent on the investigator's intuitions about what a good explanation is.

Although Woody (2004a) presents some examples of *explanatory structures*—theories, parts of theories, pictures, diagrams and other structures that play important roles in explanations—which she identified in a general chemistry textbook (Mahan and Myers 1987), she does not tell us *how* she identified these. This is not necessarily a problem for Woody, as one need not be able to identify explanations in textbooks in order to identify explanatory *structures* in them. (Although one of course needs to be able to argue that these structures are actually used in explanations *somewhere*.) But it does mean that it is necessary to expand on Woody's account in order to reach an answer to the circularity problem that Woody points to.

Looking to the few other empirical studies of explanations suggests two possible strategies for identifying explanations in textbooks. One is to rely on pre-analytic intuitions on what an explanation is—regardless of whether it is good or bad, another is to use explicit criteria for—perhaps even a definition of—when a passage in a textbook can be considered an explanation. I discuss the former of these options in the next section and the latter in Sect. 4.2.

3 Studying Science Teachers' Explanations

Zoubeida Dagher and George Cossman have categorized explanations given by science teachers in junior high schools, based on extensive empirical material (Dagher and Cossman 1992). In an earlier article Dagher (1991) provides insight into how these explanations were identified and classified. Dagher notes that identifying explanations in sources (in her case recordings of classroom discourse) can be difficult:

> While the purpose of the analysis was perfectly clear, the question about what constituted an explanation, particularly a teacher explanation, became more obscure. [...] The literature that was reviewed presented serious dilemmas. In the case of educational theory, the adoption of any particular definition appeared to fail to discriminate between 'explanations' and other categories of verbal behavior. In the case of philosophy of science, definitions tended to restrict the sense of explanation so as to eliminate instances that seemed to be legitimate teacher explanations. (Dagher 1991, pp. 68–69)

So instead of combing the transcripts with a definition Dagher adopted a grounded-theory approach in which she first searched the transcripts for passages that intuitively "looked like" explanations believing that it was possible to justify the selection later on (p. 69). When personal intuitions were unclear Dagher resorted to the "conscious and tacit entertainment of various literature based 'attributes' of explanations" (p. 70). By merging personal intuitive notions of explanation with literature based attributes, gradually a set of 'filtering lenses' emerged (p. 70), and in a post hoc formalization of definitions, guidelines were formalized that constituted a broad orientation for looking for explanations in the transcripts.

For a researcher who is philosophically minded and who knows the field under study very well this approach is likely to be productive. However, for the purposes of this study two concerns can be raised.

First of all this approach explicitly identifies what the investigator deems explanatory, and unless the investigator is highly familiar with the discipline under study this may differ from what practicing scientists in that discipline deem explanatory. This could be tested by asking practicing scientists if they agree with what the investigator has identified. But if this step is needed in order to get a useful result, why not go all the way and simply leave it to the scientists to identify the explanations in the sources? (see Sect. 4.1).

Secondly, Dagher admits that the results of her investigation would probably look different if the analysis was performed by someone else (Dagher 1991, p. 76). This is of course often the case with such interpretive studies, and it is not necessarily a problem, especially if it is mainly the finer details in the conclusions that depend on who performed the interpretive study. However, as we saw in Sect. 2.1 part of the reason why, at least philosophers, quarrel so much about explanations is that the differences in intuitions about explanations among philosophers are rather substantial. This indicates that it may not just be the finer details, but the entire outcome of the study that becomes dependent on who performs the study if it relies heavily on the intuitions of the investigator. This is certainly something one should aim to avoid to the extent possible. At least it should be made as transparent as possible how the intuitions of the investigator affected the outcome of the study.

For these reasons it will be preferable to base the study on explicit criteria that can be judged by others, or alternatively to base it on the intuitions of the practicing scientists themselves.

4 Identifying Explanations in Textbooks

The above discussion shows that one way to expose possible qualitative differences in exemplar explanations from different disciplines is to use a varied selection of textbooks as sources of explanations. When explanations in these sources have been identified the explanations from the different disciplines can be characterized and compared. While these later stages present their own challenges, my main focus here is how to identify explanations in the selected textbooks.

I have argued that it is important to explicate how this is done in order to construct a strong empirical argument. Furthermore I have argued that basing the identification largely on the investigator's intuitions will lead to results that are investigator dependent to an extent that is undesirable.

I will therefore proceed to discuss two different (but not mutually exclusive) ways to identify explanations in textbooks:

1. Ask practicing scientists from the discipline under investigation to go through the texts and identify explanations.
2. Make explicit assumptions about reliable indicators of explanations in textbooks, use these to identify the explanations in the textbooks.

I will argue in Sect. 4.1 that the first option could provide some very interesting insights if combined with follow up interviews, but that it is more suited for providing a detailed account of explanations within one specific discipline than for mapping differences in explanations from different disciplines.

In Sect. 4.2 I will discuss the theoretical advantages and limitations of the second option and provide some insight into the practical challenges as I discuss the results of a pilot study based on textbooks from chemistry and physics.

4.1 Identification by Practitioners

One way to investigate what practicing scientists deem to be explanations in a selection of textbook material could be to ask practicing scientists themselves to identify explanations in the material. This approach certainly has advantages. First of all, it ensures that the explanations found are indeed deemed to be explanations by practicing scientists, not just by philosophers or other outsiders. As I have argued, this satisfaction is not trivial to obtain through other means. Secondly, the investigator can avoid making assumptions about the nature of explanations, which might be desirable for investigators that are worried about Woody's circularity objection.

The downside to this approach is that it is likely to be very resource consuming and thus difficult to carry out in practice. Furthermore, since disagreements between the scientists about which passages in the texts are explanations are to be expected, it will be necessary to develop a way to decide when enough participants have marked a passage as an explanation to qualify as *widely accepted* as an explanation. If the requirement is that *everyone* approached has to have marked a specific passage as an explanation before it can be allowed into the pool of data then there is likely to be little data unless the number of sources that the science practitioners have to study is very high. Relying on a simple majority will on the other hand be too permissible since the minority might contain a significant number of the most experienced teachers or specialized researchers from the area of the discipline from which the candidate explanation originates.

One way to overcome these difficulties would be to gain more knowledge about the researchers (and about their position in their field) and also about why they chose the different passages for example by interviewing the practitioners afterwards or inviting them to "think aloud" while identifying explanations in textbook samples. Adding this extra layer to the investigation could yield a more detailed picture of explanations in the disciplines under investigation but is also likely to be highly time and resource consuming. Furthermore, the more detailed picture of explanations in the disciplines to be compared that this approach may yield is not strictly necessary for my current purposes. If there are significant differences across scientific disciplines then these differences are the ones that will be most relevant both in a philosophical and educational perspective, and these should be detectable through a comparison of a less detailed picture of explanations from the disciplines compared.

I will therefore go on to discuss the possibility of identifying explanations using a set of reliable indicators in order to assess whether the theoretical and practical limitations of this approach are more suited to the purposes of the current study.

4.2 Identification by Text Indicators

As argued in Sect. 2.1, a descriptive study of exemplar explanations does not necessarily result in a vicious circle if it relies on reliable indicators or even a definition of what constitutes an explanation to identify explanations in the textbooks. What I have not yet considered is which definition or indicators to rely on and whether it is practically convenient to proceed in this way.

Considering the question about indicators first, we note that explanations may be defined with reference to their function, and with reference to their structure. I have already made assumptions about an essential *function* of explanations, namely that successful explanations provide understanding. So one possibility is to start from this assumption and then investigate how understanding is gained in the discipline in question.[3] The problem with this assumption is [as Salmon also noted (1998, p. 126)] that they do not help further investigations if they are kept too general. The nature of understanding and intelligibility is not better mapped than the nature of explanation, so simply assuming that one provides the other does not help. A more detailed description of what is meant by understanding and how an explanation can provide this understanding is required for this approach to prove useful. If this is done on the basis of theoretical arguments, one could end up identifying what ought

---

[3] See Chambliss (2001) for an example of a study of explanations based on assumptions about understanding.

to provide understanding from a theoretical perspective rather than identifying what practicing scientists deem to be good explanations to give to students.[4] This is exactly what I aim to avoid.

Alternatively, one might assume something about the linguistic indicators of explanations.[5] For instance, it may be possible to search for specific language structures, or even certain key words. Against this strategy, Stephen Draper has argued that there are no linguistic traits common to *all* explanations (Draper 1988). Some, but not all, will be answers to explicitly posed why- or how-questions. Many will contain the word 'because', but some will not. More generally, Draper argues that there are no words or sentence structures that can be called *necessary* for explanations. Thus "[...] a search of a transcript for their occurrence will not pick out anything like the complete set of the explanations present" (Draper 1988, p. 20). Draper does admit that the presence of a word like 'because' can be seen as a *sufficient* condition for the presence of an explanation (p. 19). So there is nothing in the theoretical arguments that prevents us from saying that a search for keywords in a textbook could yield a good *sample* of the explanations found in the text. And this is really all we need! The question now is whether the sample will be big enough to be practically useful, and whether we have reason to think that the sample will not reflect the diversity in the explanations in the textbooks because the keyword search leaves out certain important types of explanations? I will discuss the former question of sample size in detail when I present my pilot study in Sect. 5. The answer to the latter question depends on which keywords are used. Before answering this question I will therefore have to elaborate on which keywords should be used.

4.2.1 Introducing the Keywords

A text mining approach has been used by Overton (2013) who used keyword searches to assess the importance and abundance of explanations in the journal *Science*. Overton relied exclusively on versions of the words 'explain', 'explanation'

---

[4]A different kind of objection to this approach might also be raised: Even if it can be safely assumed that any good explanation will increase the reader's understanding of the explanandum, this does not mean that a good explanation is *necessary* for an increase in understanding. Thus we will be making the fallacy of affirming the consequent if we claim to have found explanations by identifying passages that increases the readers understanding. Lipton (2009) has explored other sources of understanding (for instance thought experiments), and this potential objection could be overcome by simply assuming that Lipton's list of sources of understanding is exhaustive. If a textbook passage increases the readers understanding *and* does not belong to one of the other sources of understanding on Lipton's list, it can safely be assumed that an explanation has been found.

[5]Rowan (1988) has also discussed the advantages and challenges related to the study of explanations through assumptions about either their *function* or their *structure*. She argues that if the purpose of the study is to improve teaching, then assumptions about the function of explanations is preferable, but unfortunately she does not give us any hints as to how the practical problems associated with this approach might be overcome.

and 'explicate' as keywords indicating the presence of explanations in the articles. Although 'explanation' and 'to explain' are perhaps the most obvious candidates for a list of reliable indicators for the presence of an explanation, it seems unnecessarily restrictive to rely *solely* on these, if the aim is to identify a large and diverse set of explanations.

There is a broad consensus both in philosophy and in science education that the primary function of explanations (especially those in textbooks) is to provide understanding (De Regt and Eigner 2009; Rowan 1988). So although one cannot search for passages that will provide understanding to the reader, one can search for those passages that the author(s) of a textbook has explicitly stated *should* provide understanding,[6] passages like "To understand this …" or "this helps us understand …". Thus, versions of the words 'to understand' and 'understanding' should be added to the list of keywords to be searched. Further, as mentioned earlier, Draper (1988) (among others) acknowledges that 'because' is a reliable indicator for the presence of an explanation, and therefore can also be added to the list of keywords. Finally, one can search for answers to explicit explanation seeking questions, at least if it is possible to specify more concretely the nature of such a question. When philosophers discuss explanations they often focus on answers to why-questions (Goodwin 2003; Salmon 1998; Van Fraassen 1980), but it is also widely recognized that being an answer to a why-question is not a necessary condition for being an explanation. In addition, certain how-questions are often highlighted as explanation seeking. Mechanistic explanations, for instance, have been described as answers to how-questions (Machamer et al. 2000). Furthermore, certain what-questions may also be explanation seeking. The geologist might, for instance, try to explain *what* went on in the Cambrian explosion. In general there is no reason to believe that answers to questions involving certain interrogatives can be excluded as being explanations (Draper 1988; Faye 1999).

The reason why answers to questions other than why-questions are not discussed as much by philosophers as answers to why-questions may be that it becomes less clear when answers to these kinds of questions are explanations. Whereas we can treat all relevant answers to why questions as explanations, we cannot automatically do the same with answers to questions involving other interrogatives. For how-questions it is still relatively uncontroversial that questions about how things work are explanation seeking whereas it is more unclear whether how-much-questions like "how much ascorbic acid does a normal person need per day to avoid scurvy?" are explanation seeking. For other interrogatives it becomes even more difficult to say whether a question is explanation seeking or not simply based on the wording of the question.

Thus a reasonable way to proceed would be to start out by searching for answers to why- and how-questions (excluding how-much-questions), the word "because", and all versions of the words 'explanation', 'to explain', 'to understand' and

---

[6]Bearing in mind the possible objection raised in note 4 about the possibility of other sources of understanding.

'understanding', and use the data gained in this search to sketch the characteristics of exemplar explanations in a given discipline. This could then be used to analyze answers to other kinds of questions yielding an even more detailed picture, and so on until a sufficiently detailed account that enables one to make comparisons to other disciplines is at hand.

4.2.2 Concerns About Diversity

Having identified some useful keywords for seeking explanations in texts, I will return to the question of whether one might miss important types of explanations by relying on these keywords. Since the list of keywords contains no necessary conditions for the presence of an explanation, it is difficult to argue decisively that every type of explanation will be found. However one can argue that the keyword search will detect at least as many types of explanations as other methods. Take for instance the ten categories that Dagher and Cossman present (1992, pp. 364–366), one can argue that a keyword search would identify all of these categories.

After describing the characteristics of each of the categories Dagher and Cossman present an example of each of the ten types of explanations taken from their transcripts. Half of these examples contain either a why-question or the word 'because'. Two further categories (tautological and practical explanations) are partly defined as answers to how/why questions. So in these cases a search for key words presented above in the transcripts would not only identify explanations of the same type, but it would even identify the examples presented by Dagher and Cossman.

Is there reason to believe that explanations of the remaining three types could not be found through a keyword search? Two of the remaining categories will be familiar to most readers: teleological explanations and explanations that explain through analogy. Such explanations can be and are given as answers to why-and how-questions.

Last but certainly not least an important type of explanation in textbooks appears as descriptions of *what* happens, rather than *how* things work, or *why* they happen [as Woody has also pointed out (Woody 2003, p. 23)]. Dagher calls these genetic explanations, and they present quite a challenge for anyone who wants to distinguish between explanatory and non-explanatory descriptions. Explanatory descriptions that are also explanations are considered to provide (genuine) understanding, and therefore should be identifiable through a search for the term 'understanding' or 'to understand', given that these terms are in fact used regularly in the textbooks. As we shall see in the following section this is in fact the case, at least in introductory textbooks. All in all, this shows that the sample of explanations found through a keyword search as outlined above will be at least as diverse as a sample gained through an intuition based search.

In closing, I shall present a pilot study that illustrates how the approach outlined so far could be used in practice.

5 Pilot Study

To test the practical limitations of the first steps in the key word based approach outlined above I performed a small pilot study. I chose to focus on thermodynamics. This topic is central to both physics and chemistry, and there is an abundance of textbooks on the market aimed at audiences ranging from novices to experts. I chose a textbook from each end of this spectrum to see whether the usefulness of my approach depended on the intended audience of the textbooks. More precisely the sample studied consisted of Chaps. 17–20 (both included, 126 pages in total) from *University Physics* (Young and Freedman 2010) which is a very widely used introductory textbook in physics and Chaps. 2–8 (107 pages) from an older textbook called *Chemical Thermodynamics* (Kirkwood and Oppenheim 1961) which is "intended to serve as the basis of a senior or graduate course" for chemists.[7]

5.1 Results from the Keyword Search

I searched for each of the keywords ('because, answers to why- and how-questions plus versions of 'explanation', 'to explain', 'to understand' and 'understanding') in turn and will comment briefly on the results in the following sections.

5.1.1 Versions of 'Explanation' and 'to Explain'

I first searched the sample for versions of the keywords 'explanation' and 'to explain'. The final count of instances of either of these words in the body text of both samples was only *five*. Of the five instances two appeared in *Chemical Thermodynamics*, one in a general introduction to a chapter pointing to specific discussions later on and the other in this later discussion (Kirkwood and Oppenheim 1961, Sect. 5.2). In *University Physics* the search yielded three instances in total. Like in *Chemical Thermodynamics* one instance was in the introduction to a chapter pointing to a discussion later on (Young and Freedman 2010, Chap. 20).

Another instance was partly stated in a caption to a picture.[8] We are told that "[e]vaporative cooling explains why you feel cold when you first step out of a swimming pool" (Young and Freedman 2010, p. 568) and then pointed to a picture of three children in a swimming pool. The caption elaborates a bit on the claim made in the main text:

> [...] it may be a hot day, but these children will be cold when they step out of the swimming pool. That's because as water evaporates from their skin, it removes the heat of vaporization from their bodies (Young and Freedman 2010, p. 568)

---

[7]The book was recommended to me by a lecturer in physical chemistry as the most rigorous presentation of chemical thermodynamics that he knew of.

[8]The final instance also appears in the caption to a picture (Young and Freedman 2010, p. 564).

A striking feature of all the discussions linked to this word search is that they are not based on mathematical calculations, but rather on qualitative arguments. To the extent that they do appeal to any general laws these are postulated rather than derived. This is particularly striking for *Chemical Thermodynamics*, since it aims to "present a rigorous and logical discussion of the fundamentals of thermodynamics [...]" (p. v).

5.1.2 How-Questions

Learning how is apparently important in introductory physics. Each of the chapters from *University Physics* states the learning goals of the individual chapter. The three chapters in the sample state a total of 27 different things that the student should understand after reading the chapters. 21 of these contain the word 'how', none of them the word 'why'.

Four main categories of how-questions were identified covering most, but not all instances: (1) how-much-questions, asking for the value of a particular variable under specific conditions, (2) questions about how some natural or artificial thing works—e.g. how different kinds of thermometers work, (3) how concepts or laws are related—e.g. the relation between Newton's laws and the ideal gas law—and finally (4) how to answer a how-much-question given specific conditions.

As discussed in Sect. 4.2.1, answers to questions of type (1) cannot immediately be taken to be explanations, whereas I find it safe to assume that answers to the remaining types of how-questions are explanations when found in textbooks. How-much-questions are by far the most abundant in *University Physics*, as the exercises following each chapter have a very high proportion of how-much-questions, and most occurrences of 'how' are found here. In the main text around half of the how-questions are how-much-questions. Based on a further analysis of the answers provided to why- and other types of how-questions it is quite possible that some of the answers these how-much-questions may be identified as explanations as well. Focusing for now on the more unambiguously explanation seeking how-questions—there are roughly 5–10 per chapter—one finds that the answers are often quite detailed and relatively technical, taking up more space than for instance answers to why-questions.

Since the vast majority of how-questions in *University Physics* occur in the exercises, and since there are no exercises in *Chemical Thermodynamics,* one would expect there to be much fewer how-questions in this more advanced textbook.[9] Indeed the search through *Chemical Thermodynamics* yielded only three instances. All three instances were of type (3) described above about how concepts

---

[9]Furthermore, the physical format of the two books is quite different, so the number of words on a page with no equations or figures is about 50 % higher in *University Physics* than in *Chemical Thermodynamics*. Thus even if the key words were equally frequent in the two texts I would still have identified a more instances in *University Physics* than in *Chemical Thermodynamics*. One should thus be careful not to read too much into the absolute differences in the number of instances of any of the individual keywords between the two texts.

can be related. Two of the instances figure prominently as the framing questions for an entire chapter and thus indicates that understanding how is important in advanced chemistry as well.

5.1.3 Why-Questions

The word 'why' appears 992 times in the whole of *University Physics*,[10] so even though Woody might be right that there is lots of explanatory content that is not phrased as answers to specific questions (Woody 2003, p. 23) it might turn out that there is simply so much explanatory content in these books that the small fraction of it that *is* phrased as direct answers to explanation seeking questions will be more than enough for the purposes of this method. In the sample chapters from *University Physics* the word 'why' appears between five and seven times per chapter. Although this means that 'why' is roughly as abundant as the word 'how' when used in an explanation seeking context, the answers to why-questions are not as detailed and do not take up as much space. The same is true for *Chemical Thermodynamics* where the word 'why' appears only once in the sample in order to point to a phenomenon that thermodynamics *cannot* explain.

It is interesting that explanation seeking how-questions are so much more prominent in both textbooks compared to why-questions since it challenges the approximation commonly used in philosophy that all explanations are answers to why-questions. If further studies can establish that why-questions do not play a significant role in advanced textbooks then perhaps the appropriateness of approximating explanations with answers to why-questions should be re-evaluated.

5.1.4 Because

The word "because" is abundant in both texts. A prominent feature of the explanations involving "because" in both samples is their qualitative nature and brevity. Often the explanandum and the explanans are contained within a single or just a few sentences. Take for example the following passage:

> The thermal conductivity of "dead" (that is non-moving) air is very small. A wool sweater keeps you warm because it traps air between the fibers. In fact, many insulating materials, such as Styrofoam and fiberglass are mostly dead air. (Young and Freedman 2010, p. 571)

In this respect they resemble the explanations found in the search for "explanation" and "to explain".

---

[10]This makes it more abundant than the words 'explain' and 'understand' which appear 821 and 500 times respectively, but less abundant that the word 'because' which appears 1009 times. The word 'how' occurs 2140 times but as mentioned the majority of these appear in how-much-questions.

5.1.5 Understanding

The search for the versions of 'understanding' and 'to understand' in *Chemical Thermodynamics* yielded only two occurrences. One coincided with an instance of because and one coincided with the single occurrence of "why" pointing to the lack of understanding of the expressions for the entropy of a gas (real or ideal) until the advent of quantum mechanics.

University Physics yielded 55 hits. Versions of the words regularly show up in the introduction to chapters or sections to debut the theory that will be explained later or to stress the importance of certain explanations. For instance the following statement occurs after a passage describing the temperature dependence of the internal energy, U, of an ideal gas (T denotes the temperature):

> Make sure you understand that U depends only on T for an ideal gas, we will make frequent use of this fact (Young and Freedman 2010, p. 636)

Thus the preceding argument is meant to provide understanding to the student, and should be considered an explanation. Understanding is commonly used in *University Physics* to point to other passages that for the purposes of this study can be treated as explanations. It is not always clear, however, how the promised understanding will be provided. For instance the following is found in the introduction to Chap. 17:

> The concepts in this chapter will help you understand the basic physics of keeping warm and cool. (Young and Freedman 2010, p. 551)

In this instance further indicators are needed about *where* in the chapter this understanding is provided and *how* it is provided. A partial answer to this question is that the concepts help us to answers to certain why-questions like the one concerning children in a swimming pool mentioned above (Sect. 5.1.1).

Most instances of the word "understanding" appear in the titles of the 'Test your understanding' questions that are generously distributed throughout the whole text.[11] These explanation seeking questions allow the student to test whether she has gained sufficient understanding of the subject matter discussed in the preceding section to proceed to the next sections. The authors' answers to these questions are given at the end of each chapter. These answers could be relevant for the current study since they explicitly serve as guides to what an appropriate answer to an explanation seeking question looks like in the current discipline at the current level.

This leads to the more general question of how one can use the exercises in textbooks as sources of explanations.

---

[11]This type of questions is common in more recent introductory textbooks.

5.2 Including the Exercises

Learning to construct high quality explanations requires practice. In science education an important part of this practice comes through solving textbook exercises and evaluating the answers. Explanations are among the kinds of answers the textbook question writer is hoping to elicit.

I will refer to an exercise that is formulated as an explicit request for an explanation or as an explanation seeking how- or why-question, as an 'Explanation Requesting Exercise' or just an ERE. Could one extend the material searched to include the exercises in order to identify the EREs and perhaps use the solution manuals containing elaborated solutions to exercises that are available for many textbooks as a source of explanations?

I believe that this approach could be useful, but it is important to be sensitive to its limitations. Introductory textbooks usually contain an abundance of exercises. *University Physics* for instance contains well over a hundred exercises after each chapter, and though the majority of the exercises are not EREs the sheer number of exercises means that it will be possible to get some data. The more advanced textbooks generally contain much fewer exercises than introductory textbooks and the density of EREs is also much lower.

5.2.1 University Physics

The EREs in *University Physics* fall into two general groups. The first group (a) is formulated as a description of a phenomenon or a result of a calculation combined with a request for an explanation. For instance (Young and Freedman 2010, p. 622):

> Explain why in a gas of N molecules, the number of molecules having speeds in the finite interval $v + \Delta v$ is $\Delta N = N \int_{v}^{v+\Delta v} f(v)dv$

Answers to this type of exercise can be taken as explanations without question.

The second group (b) contains EREs where students are asked to explain their reasoning behind a certain answer. This kind of ERE is completely absent from the more advanced textbook passages that I have looked at (see next section). In the type (b) exercises the main function of the word explain is to force the students to make an elaborate answer. Take for instance the following discussion question (Young and Freedman 2010, p. 579):

> Q18.23: If the root-mean-square speed of the atoms in an ideal gas is to be doubled, by what factor must the Kelvin temperature of the gas be increased? Explain

The question posed can be answered by stating a single number ("4"). Thus if "explain" was omitted it could easily be thought that a satisfactory answer to this question is simply "4". However the addition of the word explain indicates that the student has to come up with a more elaborate answer such as "4, because the root-mean-square speed of the atoms in an ideal gas is proportional to the *square root* of the Kelvin temperature of the gas".

"Why/Why not" or simply "Why?" is also often added after questions that can be answered very briefly. Thus the addition of "Why?" or "Explain" after other kinds of questions than why-questions can be seen as a clarification from the authors that the question just posed is indeed an explanation seeking question, not just what might be called a fact seeking question.

Type (b) EREs are particularly interesting for two reasons. First they pose explanation seeking questions that are not why-questions. As mentioned there is some consensus that answers to this type of question should be treated as explanations, but little attention has been given to them so far. Secondly one will need knowledge about this kind of explanation seeking questions if data is to be gathered from the exercises in the more advanced textbooks,[12] since EREs are so rare in these, as I will illustrate below.

5.2.2 More Advanced Textbooks

Chemical Thermodynamics contains no exercises at all. I therefore made a brief search in two physics textbooks, *Introduction to Electrodynamics* (Griffiths 1999) and *Statistical Physics* (Mandl 1988), both aimed at slightly more advanced physics students than *University Physics*. Searching the exercises of two random chapters in each book[13] gave a total of 65 exercises none of which contained versions of the word 'explanation' or 'to explain'. The search for explanation seeking how- and why-questions yielded only three results, all from *Introduction to Electrodynamics*. Most exercises in these two books are formulated not as a question, but as a request to "find", "show" or "calculate" something. This indicates that even if an elaborated solution manual can be found to these more advanced textbooks, it may still be difficult to use the keywords discussed here, to find explanations in them.

A further complication related to the more advanced textbooks is that it is usually only the introductory textbooks that have solution manuals that elaborate on how the exercises should be solved. Most solution guides just give the result, and even the most elaborate solution manuals do not always contain solutions to the EREs, especially if the solution requires the construction of a qualitative argument. Thus the inclusion of the exercises from the more advanced textbooks might not be of much use in practice if due to time limitations it is not possible to make large scale studies of how practitioners would solve these EREs, or unless the investigator herself is capable of producing the solutions, which would require something

---

[12]Assuming that there is data to be found. It may be that the reason why explanation seeking questions are so hard to find in more advanced textbooks is because they are not posed, but given the commonness of explanation seeking questions in everyday discourse and the consensus among philosophers and scientists that explanations are important in science I find that highly unlikely.

[13]Chapters 3–4 in Mandl (1988) and Chaps. 8–9 in Griffiths (1999).

close to contributory expertise (as opposed to interactional expertise (Collins 2004)) for the most advanced textbooks.

All in all I find that the exercises from the introductory textbooks and their solutions could be a valuable source of explanations from the different disciplines. However it is not yet clear whether the exercises from more advanced textbooks can become as valuable, since the tools discussed in this chapter are of limited use when trying to analyze these exercises. However the results gained through the analysis of the introductory textbook exercises could provide the necessary tools for studying the more advanced exercises.

6 Conclusion

Studies of scientific explanations by philosophers have served as background and inspiration for studies in science education on many occasions. Although it is widely recognized that explanatory practices differ between scientific disciplines, philosophers have almost exclusively focused on what similarities there may be. The aim of this chapter was to develop an empirical method for exposing possible differences in exemplar explanations given by practicing scientists to students. Empirical studies of explanations are rare in both philosophy of science and science education. The ones that exist share the assumption that the nature of explanations is best studied through the identification of a set of concrete explanations that can be used as the basis of an inductive argument. I have followed this assumption in this chapter, and thus I have not considered other possible approaches to an empirical study of explanations.[14]

When presenting an empirical study of explanations based on a set of concrete explanations it is important that the data collection procedure is made transparent. Since it is not essential to identify every explanation in a textbook in order to have a useful data set, and since there is no consensus on a definition of an explanation, I argued that one useful approach is to use a set of reliable linguistic indicators as the basis for gathering explanations from science textbooks. The list consisted of a number of keywords that should be fairly uncontroversial to use as identifiers of explanations. I then showed that the use of these keywords could yield a dataset that was as varied as the dataset gathered by Dagher, and the pilot study showed that the approach yields substantial amounts of data, especially from the search for explanation seeking how-questions. The pilot study also indicated that why-questions and references to understanding are more common introductory textbooks than the advanced textbooks produced a little less data. This highlighted that it may be important either to include larger samples of advanced textbooks in the sources or

---

[14]An alternative approach would be to interview practicing scientists and ask them what characterizes good explanations for students or what they think are the main differences between the explanations from their discipline and explanations from other disciplines. How useful such an approach would be is an empirical question, and I am not aware that it has ever been attempted.

to reevaluate the search criteria after the first search, and go through the sources more than once using increasingly sophisticated criteria.

The method developed here was designed to fit a very specific purpose, and parts of the argumentation rests heavily on this specific purpose, especially the arguments for limiting the study to just textbooks. However, the usefulness of an empirical approach based on relevant sources of explanations is not dependent on the specific purposes considered here. As mentioned one could gain a very detailed picture of explanations in any scientific discipline if a combination of textbook studies and interviews was conducted. Such studies could provide an important supplement to the many case studies of explanations from different disciplines that are in the current philosophical literature.

Acknowledgements This chapter benefitted from critical feedback from Hanne Andersen, Douglas Allchin and an anonymous reviewer. Furthermore it benefitted from insights and comments provided by the entire philosophy of contemporary science in practice group at Aarhus University. The group is supported by the Danish Council for Independent Research|Humanities.

References

Braaten, M., Windschitl, M.: Working toward a stronger conceptualization of scientific explanation for science education. Sci. Educ. **95**, 639–669 (2011)

Chambliss, M.: Analyzing science textbook materials to determine how "persuasive" they are. Theory Pract. **40**, 255–264 (2001)

Collins, H.: Interactional expertise as a third kind of knowledge. Phenomenol. Cogn. Sci. **3**, 125–143 (2004)

Dagher, Z.: Methodological decisions in interpretive research: the case of teacher explanations. In: Gallagher, J. (ed.) Interpretive Research in Science Education, pp. 61–83. National Association for Research in Science Teaching (1991)

Dagher, Z., Cossman, G.: Verbal explanations given by science teachers: their nature and implications. J. Res. Sci. Teach. **29**, 361–374 (1992)

De Regt, H.W., Leonelli, S., Eigner, K.: Scientific understanding: philosophical perspectives. University of Pittsburgh Press, Pittsburgh (2009)

Draper, S.: What's going on in everyday explanations. In: Antaki, C. (ed.) Analysing Everyday Explanation: A Casebook of Methods, pp. 15–32. Sage Publications, London (1988)

Edgington, J.R., Barufaldi, J.P.: How research physicists and high-school physics teachers deal with the scientific explanation of a physical phenomenon. Paper Presented at the Annual Meeting of the National Association for Research in Science Teaching, San Francisco, CA, 22–25 Apr 1995

Faye, J.: Explanation explained. Synthese **120**, 61–75 (1999)

Friedman, M.: Explanation and scientific understanding. J. Philos. **71**, 5–19 (1974)

Goddiksen, M.: Interdisciplinarity and explanation: how interdisciplinary education calls for a new approach to research on scientific explanations. In: Proceedings of the 12th Biennial IHPST Conference. http://archive.ihpst.net/2013-pittsburgh/conference-proceedings/ (2013)

Godfrey-Smith, P.: Theory and Reality: An Introduction to the Philosophy of Science. University of Chicago Press, Chicago (2003)

Goodwin, W.: Explanation in organic chemistry. Ann. N. Y. Acad. Sci. **988**, 141–153 (2003)

Griffiths, D.J.: Introduction to Electrodynamics, 3rd edn. Prentice Hall, Upper Saddle River (1999)

Hempel, C.G.: Aspects of Scientific Explanation. The Free Press, New York (1965)

Kirkwood, J.G., Oppenheim, I.: Chemical Thermodynamics. McGraw-Hill, New York (1961)
Kuhn, T.S.: The structure of Scientific Revolutions, 3rd edn. University of Chicago Press, Chicago (1996)
Lipton, P.: Understanding without explanation. In: de Regt, H.W., Leonelli, S., Eigner, K. (eds.) Scientific Understanding, pp. 43–64. University of Pittsburgh Press, Pittsburgh (2009)
Machamer, P., Darden, L., Craver, C.: Thinking about mechanisms. Philos. Sci. **67**, 1–25 (2000)
Mahan, B.M., Myers, R.J.: University Chemistry, 4th edn. Benjamin/Cummings cop, Menlo Park, Calif (1987)
Mandl, F.: Statistical Physics, 2nd edn. Wiley, Chichester (1988)
McMullin, E.: Rationality and paradigm change in science. In: Horwich, P. (ed.) World changes: Thomas Kuhn and the Nature of Science. MIT Press, Cambridge (1993)
Nersessian, N.J.: Opening the black box: cognitive science and history of science. Osiris **10**, 194–211 (1995)
Ogborn, J., Kress, G., Martins, I., McGillicuddy, K.: Explaining Science in the Classroom. Open University Press, London (1996)
Overton, J.A.: "Explain" in scientific discourse. Synthese **190**, 1383–1405 (2013)
Peker, D., Wallace, C.: Characterizing high school students' written explanations in biology laboratories. Res. Sci. Educ. **41**, 169–191 (2011)
Rowan, K.E.: A contemporary theory of explanatory writing. Written Commun. **5**, 23–56 (1988)
Salmon, W.C.: Why ask "why?"? An inquiry concerning scientific explanation. In: Salmon, W. (ed.) Causality and Explanation, pp. 125–142. Oxford University Press, New York (1998)
Solomon, J.: Higher level understanding of the nature of science. Sch. Sci. Rev. **76**, 15–22 (1995)
Treagust, D., Harrison, A.: The genesis of effective scientific explanations for the classroom. In: Loughran, J.L. (ed.) Researching Teaching: Methodologies and Practices for Understanding Pedagogy, pp. 28–43. Routledge, London (1999)
Unsworth, L.: Evaluating the language of different types of explanations in junior high school science texts. Int. J. Sci. Educ. **23**, 585–609 (2001)
Van Fraassen, B.C.: The Scientific Image. Clarendon, Oxford (1980)
Woodward, J.: Scientific explanation. In: Zalta, E.N. (ed.) Stanford Encyclopedia of Philosophy. http://plato.stanford.edu/archives/win2011/entries/scientific-explanation/. Accessed 4 May 2014 (2011)
Woody, A.I.: On explanatory practice and disciplinary identity. Ann. N. Y. Acad. Sci. **988**, 22–29 (2003)
Woody, A.I.: Telltale signs: what common explanatory strategies in chemistry reveal about explanation itself. Found. Chem. **6**, 13–43 (2004a)
Woody, A.I.: More telltale signs: what attention to representation reveals about scientific explanation. Philos. Sci. **71**, 780–793 (2004b)
Young, H.D., Freedman, R.D.: Sears and Zamansky's University Physics: With Modern Physics, 13th int. edn. Addison-Wesley, San Francisco (2010)

Longino's Theory of Objectivity and Commercialized Research

Saana Jukola

Abstract In this paper, I shall examine Helen Longino's view on the objectivity of science and study how it can be applied to the evaluation of current scientific practices. By discussing two prominent cases in biomedical research, I articulate some epistemically alarming features of commercialized research and highlight the importance of paying attention to the context of scientific inquiry. In addition, I claim that the examined cases can help uncover philosophically interesting empirical work on extra-scientific mechanisms influencing research practices.

Keywords Objectivity · Helen longino · Biomedical research · Commercialization

1 Introduction

In her books *Science as Social Knowledge* (1990) and *The Fate of Knowledge* (2002), Helen Longino aims to construct a philosophical theory of science that is sensitive to the knowledge gained through sociological and historical studies of science. She states that her philosophy is a philosophy of real-life science with researchers who do work in an ideal, interest- and context-free research setting, but who have interests, values and a social context of their own:

> I insist […] on an epistemology for living science, produced by real, empirical subjects. This is an epistemology that accepts that scientific knowledge cannot be fully understood apart from its deployments in particular material, intellectual, and social context. (Longino 2002, 9).

A central part of Longino's project is a theory of the objectivity of science which does not connect objectivity with the value-neutrality of research. She has also introduced criteria for evaluating research communities, the purpose of which is to

S. Jukola (✉)
University of Jyväskylä, Jyväskylä, Finland
e-mail: saana.jukola@jyu.fi

define the features a community should fulfill in order to produce reliable knowledge. In this paper, I shall focus on Longino's views on objectivity and on how to organize research in a way that secures objectivity in Longino's sense. The aim is to complement her view, rather than to argue against it. I shall study the implications of Longino's theory by discussing two cases of biomedicine. These prominent and much discussed cases give reasons to believe that certain features of commercialized research[1] can lead into epistemically alarming consequences. I chose to use the cases of commercialized research since the commodification of research has affected universities in several ways during the last few decades, and thus, taking notice of this phenomenon is vital if one wishes to study scientific activity as it is practiced in the real world by "real, empirical subjects".

I use empirical studies and reports on research on the Selective serotonin reuptake inhibitors and so-called biological psychiatry to supplement Longino's view. The central point is the conditions for objectivity do not depend only on the internal factors of the communities of researchers. I outline how critical interactions within and between scientific communities, i.e., a necessary condition for objectivity in her sense, are dependent on factors and actors external to research activities: when current, living science is examined, it is essential to consider how funding arrangements and science policy decisions can mold the way research is conducted. The reason for invoking these cases is that studying concrete examples makes it clearer why certain extra-scientific developments affect the independence of scientific research and guide its results.

The structure of the paper is as follows: In Sect. 1, I shall outline the main features of Longino's social view on objectivity, whereas the criteria for evaluating research communities are introduced in Sect. 2. In Sect. 3, I shall focus on examining two cases of commercialized research by investigating them while keeping in mind Longino's definition of objectivity. The first example deals with individual conflicts of interests and the second with the way commercial interests may shape the research agenda. The essay ends with some remarks on the cases.

2 Longino on Objectivity

Discussing the objectivity of science, Longino focuses on the objectivity of the scientific method and in particular on the question of how a method "provides means of assessing hypotheses and theories in an unbiased manner" (Longino 1990, 63). She also wants to distinguish the objectivity of (a) scientific method from (b) the objectivity of individual researchers and their attitudes and practices. Longino claims that these two are often conflated and thus the objectivity of science

---

[1] With the commercialization of research I refer to the same phenomenon as Irzik (2010, 130): when research is commercialized, "…scientific research is done, scientific knowledge is produced, and scientific expertise mobilized…primarily for the purpose of profit".

is viewed in a highly individualistic way. Longino's stance is that "the objectivity of scientific inquiry is a consequence of this inquiry's being a social, and not individual, enterprise" (Longino 1990, 67).

Longino argues that the traditional view concerning the objectivity of science, which associates the objectivity of scientific research with the ability of individual researchers to use their sense organs and reasoning in an appropriate way, is insufficient. Examinations of scientific observation and evidential reasoning reveal the deficiencies of the individualistic view. First of all, Longino states that social negotiations play an essential role in defining what evidence is: questions regarding the boundaries of classes and concepts, and the importance of categories need to be answered before observational reports can be turned into evidence. The demand for the repeatability of experiments also highlights the social nature of observation (Longino 2002). A scientist has to face challenges to her observations and be able to defend and, when needed, modify her accounts before they are accepted.

Second, the social nature of scientific inquiry becomes evident through a scrutiny of scientific reasoning. When scientists examine the world, they do not only try to describe it on the level of sense perception, but to find principles and processes that underlie this level. In the explanations of phenomena scientists postulate processes and entities that do not occur in the descriptions of the phenomena (Longino 2002). This leads to questions concerning the relationship between evidence and hypotheses—and to the problem of under determination: Why is something taken as evidence for something else? What makes us think that a certain state of affairs supports a hypothesis we have?

According to Longino's formulation, the under determination problem boils down to the fact that there is a gap between evidence and hypotheses, as "[s]tates of affairs…do not carry labels indicating that for which they are evidence or for which they can be taken as evidence" (Longino 1990, 40). This entails the context dependency of evidential support. The other assumptions researchers have, the theory they apply and the concepts they use affect what they consider to be relevant data and their interpretations of that data. Longino (1990, 41) states, "What determines whether or not someone will take some fact or alleged fact, x, as evidence for some hypothesis, h, is not a natural (for example, causal) relation between the state of affairs x and that described by h, but are other beliefs that person has concerning the evidential connection between x and h". Inferences from evidence to hypotheses are enabled by the so called background assumptions, concerning for instance the way the world is structured, what sort of evidence is relevant and what kinds of explanations are desirable. These assumptions working as premises in the reasoning often remain implicit, and may not be recognized by researchers holding them.

A noteworthy feature of Longino's view on objectivity is that according to it, science can be objective even if it is not value-free, and striving for value-freedom of science actually might not even be a desirable goal—and this does not only mean that value neutrality in itself is a value. She makes a distinction between what she calls constitutive and contextual values. Constitutive values are values that scientists have traditionally appealed to when judging explanations: accuracy, simplicity and truth, for instance. Contextual values, on the other hand, are social and cultural values and preferences (Longino 1990).

The role of constitutive values in the practice of science is generally not questioned.[2] Often, however, the intrusion of contextual values in the research process is viewed as detrimental to objectivity. According to Longino's theory, social and cultural values can enter scientific reasoning via background assumptions, and thus function as constitutive values. This can happen, for example, when value-laden terms are used to describe data or when decisions concerning the type of data to be gathered are made (Longino 1990). How, then, is objectivity of science possible if reasoning from evidence to hypotheses is context-dependent? Longino's answer is that instead of pursuing the unattainable goal of value-freedom, we have to keep the values and assumptions steering our knowledge productive practices in check. Contextual values do affect scientific research, and "whether [the effect] is positive or negative depends on our orientation to the particular values in question" (Longino 1990, 218). Because of this, the importance of questioning the assumptions that steer research is highlighted by Longino: we cannot get rid of the values completely, but we can discuss which values we can accept as affecting our science. As an example of this, she mentions the "man-the-hunter" and "woman-the-gatherer" perspectives on human descent. Both of these theories were given as answers to the question of how changes in the habit of tool use affected the evolution of the species. The "man-the-hunter" perspective, which had been promoted for example by Edward O. Wilson and Sherwood Washburn, focused on changes in the hunting behavior of males and the use of stone tools while the "woman-the-gatherer" perspective (Nancy Tanner and Adrienne Zihlman supported of this view, among others) assigned a major role to females and use of tools made of organic material, such as sticks. Both of these theories explained the same data and were gender-centered. Since no direct evidence for neither of the approaches was available, the choice between them could not be made on the basis of empirical evidence alone (Longino 1990).

By Longino's account, the gap between data and hypotheses and the role of background assumptions in reasoning are only problematic if we hold that scientific inquiry is an individual effort. When critical interactions within and between research communities are taken to be essential parts of the scientific method, we can see how the influence of biases can be regulated even when background assumptions play a role in reasoning. The objectivity of scientific research, i.e., the independence of accepted hypotheses from individual biases, cannot be secured without subjecting evidence, reasoning and results to the critical scrutiny of the community. Scientific method is something that is practiced by groups, not by individuals (Longino 1990). Because the background assumptions work as premises in

---

[2]In 1995 Longino questions her original distinction between constitutive and contextual values and argues that what we treat as paradigmatic constitutive values is partly dependent on our contextual values.

reasoning, but may often not be recognized by researchers holding them, it is essential that the reasoning process can be checked by other members of the community.

According to Longino's theory, it is important to encourage criticism that enables one to question the soundness of background assumptions. Even if these assumptions affect reasoning, their role can be uncovered and they can be criticized, defended and modified by critical examination. Conceptual considerations are an essential element in scientific enquiry and the method of science has to include not only hypotheses testing "through comparison with experiential data" but also "subjection of hypotheses and background assumptions in light of which they seem to be supported by data to varieties of conceptual criticism" (Longino 1990, 74). In other words, it is particularly important to examine why certain data is thought to be evidential for a certain hypothesis at all. Without this type of criticism the objectivity of research could not be ensured. Publicity of science is a necessary condition for criticism, and thus, for objectivity as well (Longino 1990). However, publicity alone is not a sufficient condition but requires different points of view that enable criticizing the accepted beliefs.

It should be mentioned that even though Longino highlights the importance of conceptual considerations, she does not disregard the role empirical evidence plays in the evaluation of theories. Nevertheless, since the validity of background assumptions cannot always be empirically tested, the objectivity of research should not be associated with the empirical features of inquiry alone (Longino 1990).

3 Longino's Criteria for Objective Communities

As I hope is clear at this point, Longino sees the critical interactions within and between research communities as important in knowledge production, in that without them, the objectivity of research could not be guaranteed. Next I shall introduce the criteria she offers as tools for the evaluation of communities: fulfilling these criteria denotes that discursive interactions in a community can be labeled as effective and the community as objective (Longino 1990, 2002):

1. *Venues*. In order to be named as objective, the community must have publicly recognized forums where evidence, methods, reasoning and assumptions can be criticized. Journals, conferences and the peer review process are examples of these. In addition, giving criticism and responding to it should be valued as highly as presenting original research, and criticism should be presented and evaluated according to the same standards. This is because questioning the views previously taken for granted is necessary for the evaluation of whether the established views are tenable as well as in bringing forth new ways of understanding the subject.
2. *Uptake of criticism*. This criterion requires that beliefs and theories of the community must change in response to criticism. Members of the community

need to pay attention to criticism and be willing to respond to it, not just tolerate dissent. Also critics must take notice of the responses they receive. Longino states that the individual members of the communities are not bound to change their beliefs, but the communities must be ready to take criticism into account.

3. *Shared standards.* Members of the community must have some shared and publicly recognized standards, by reference to which hypotheses, observational practices and theories can be evaluated. Shared standards are necessary for the participants in critical discussions to recognize where they agree and where they disagree. If criticism does not appeal to anything that is agreed upon by those whose position is criticized, it is not relevant to the position. These standards are derived from the goals of the community and can be epistemic or social values (e.g., truth, empirical adequacy, consistency with accepted theories or relevance to social needs) or substantive principles. The standards shared by the community must themselves be sensitive to criticism and subordinated to the cognitive goals of the community.

4. *Tempered equality.* The purpose of the criterion is to disqualify those communities where irrelevant factors, such as the political, social or economic power of individuals or groups, have an effect on which assumptions are accepted. No point of view should be excluded from discussion unless it fails to fulfill certain standards decided collectively by the community to determine which positions are qualified and which are not. Equality should be tempered because the intellectual capacities of individuals differ: It would not be reasonable to require that all members of the community should have their say in all matters, no matter what their qualifications were. The third criterion, shared standards, also sets limits to the requirements for the participants in the discussion. Those who have accepted the standards should give and take criticism, but the standards limit the scope of criticism to which the community must respond to only those statements with an effect on satisfying the goals of the community. If a discussant recurrently appeals to something that is not accepted by the community or repeats the same complaints all over again without taking notice of the responses, she loses her status as an equal member of the discursive community.

By utilizing the above norms, we should be able to recognize which social interactions are knowledge productive and which are not; for Longino these criteria assure the objectivity of research (Longino 2002). These criteria are the "features of an idealized epistemic community" (Longino 2002, 134), and when they are followed, the influence of biasing factors on accepted views can be kept in check, even if individuals behaved in a way that would not be qualified as objective by traditional standards. In other words, the criteria are meant to secure the publicity of research and cultivation of critical points of view. Allowing inter-subjective criticism is not enough, but the community must be responsive to it and encourage the questioning of adopted views.

4 Longino's View on Objectivity and Commercial Research

Next I shall contrast Longino's view on objectivity with two cases of commercialized biomedical research. Remember that according to Longino, what objectivity boils down to is that hypotheses and theories can be assessed in an unbiased manner. I take that being unbiased, correspondingly, denotes that the preferences of researchers or other involved parties have not unduly steered the process towards certain types of outcomes (Jukola forthcoming; Wilholt 2009). The following short examples demonstrate how research can fail to be objective in this sense. In addition, they help us supplement Longino's criteria and focus our attention on some features of current research that can turn out to be particularly valuable for the philosophy of science. However, before introducing the cases, I shall briefly take a look at some of the comments Longino's views on objectivity have attracted before.

To begin with critical remarks, Longino's criteria have been criticized by Barwell (1994), Biddle (2007), Leuschner (2012) and Smith (2004), among others. One of the detected problems with the criteria is that they are meant to be a suitable tool for the evaluation of all scientific communities, but, as Biddle (2007) states, it is unlikely that the epistemic issues facing all disciplines are similar. Fields of pure or applied research, or highly commercialized or politically relevant fields or areas of research without foreseeable applications may all contain different dubious features that threaten objectivity.

The criteria have also been denounced as too vague (e.g., Biddle 2007, 33; Smith 2004, 145–146): What does it mean that a community should "cultivate all relevant perspectives"? What does it mean that financial interests should not have a role to play in judging hypotheses? Who decides the right credentials for entering critical discussions concerning a certain matter—is a university degree needed or are laypeople admitted? According to Leuschner, this is a problem because Longino's criteria (3) and (4) imply circularity, as the pluralism of perspectives is called for and at the same time certain standards are needed for limiting the cacophony of unqualified voices. Thus emerges the question of who decides upon the adequate credentials for entering the community (Leuschner 2012).

Longino's goal is to formulate a philosophical theory of science that is sensitive to what is happening in the real world. I aim at offering some answers to questions that are left unanswered by the criteria: Which mechanisms can hinder the development of relevant alternative perspectives? In what ways may financial interests play a role in the evaluation of hypotheses? I want to examine these questions by focusing on certain features of the current, commercialized research culture. I shall argue that the diversity Longino sees as necessary for achieving objectivity can be threatened by institutional practices that one could call extra-scientific, i.e., not used by those who are "professionally engaged in scientific research" (Longino 1990, 69, n 10). Thus, the examples highlight the need of paying attention to the so-called "discovery side" of science (c.f. Brown 2010). In addition, the examples presented below support Longino's view on the importance of community-level processes. Namely, even if individual researchers were testing hypotheses in a manner advocated by the promoters of more individualistic views on objectivity, that is, by basing their conclusions "on strict logical adherence to relevant facts" (Smith 2004, 152), the outcomes of research may still unjustifiably reflect the preferences of involved parties, such as funders. This is because biases may enter the process at a stage when research questions and hypotheses are generated, and thus, steer inquiry towards certain kinds of results even before the testing of different claims has begun.

Although I use cases of commercialized research as examples, it is not my intention to lament the changing norms of academic science and yearn for the golden age of "pure science". Collaboration between academia and private firms has beneficial consequences for the advancement of science as well.[3] Research conducted in search for new commercializable technologies can lead to theoretical discoveries, or "application innovations" as Carrier (2010, 172) calls them. With the financing received from private industry, universities can supplement public funding in order to, for example, hire staff, conduct research and hold scientific conferences (Shamoo and Resnik 2009). The goal of this paper is to identify some mechanisms through which commercialization affects knowledge productive practices in a way that potentially restrains the diversity of research communities and the conditions for critical interaction, not to denounce the collaboration between academia and industry altogether.

4.1 Conflicts of Interests and Study Design

I begin with individual conflicts of interest (COIs). In discussing such conflicts, I shall follow the definition given by Shamoo and Resnik (2009, 191): "An individual has a conflict of interest when he or she has personal, financial, professional, or political interests that are likely to undermine his or her ability to meet or fulfill his or her primary professional, ethical, or legal obligations".[4] In the context of science, COIs can arise, for instance, when researchers receive stock in the pharmaceutical company the products of which they are testing or when the funder of their project urges them to refrain from publishing certain results (Shamoo and Resnik 2009, see also Krimsky 2003). In the academic world, COIs can also result from the need to secure funding. Future funding for a project can be dependent on achieving positive results. If negative results are considered to equal no results, researchers may face the temptation to choose their methods and data in a way that ensures positive results. The research funded by public money is not, however, immune to this problem either. Nevertheless, I will next discuss the complications arising from a situation where abundant part of research is supported by industry.

---

[3]E.g., Shapin (2008) states that the commercialization of research may in some cases increase the freedom of researchers.

[4]I am not going to discuss the adequacy or exhaustiveness of this definition here.

Social scientific research has shown that even small financial ties can alter the abilities of an individual to fulfill her obligations. In their study, Katz et al. (2003) discussed the influence that small gifts received from pharmaceutical firms have on the prescription behavior of physicians. Their argument is that the biasing effect of free meals or even pens and notepads stems from the social norm of reciprocity: we feel the need to return favors and gifts in one way or another. In another study, Babcock et al. (1995) examined how self-serving biases operate in judicial decision-making and interpretation of information. In their experiment the subjects were appointed to play the roles of either a plaintiff or a defendant and were then given the same information about a case (a motorcyclist suing an automobile driver for the damages of an accident). After reading the case material and negotiating with the opponent, the subjects had to estimate how a neutral party ("a judge") would rate the importance of a set of arguments, half of which were supporting the plaintiff and the other half the defendant. The aim of Babcock et al. was to see if the judgments of a person were dependent on the position he or she was randomly assigned to. And indeed they were. Subjects tended to believe that a neutral party would consider arguments in favor of themselves as more important than the arguments in favor of the opponent. According to Babcock et al. (1995), this suggests that self-interest biases the way human beings process information. These studies suggest that the interests an individual has can influence the way she evaluates different claims, even if she pursues neutrality. In other words, we can be affected by COIs even without recognizing this ourselves. In the context of scientific research, extraneous interests may influence study design, collection and analysis of data, and publication (Schafer 2004). This supports Longino's claim that we ought not to reduce the objectivity of science to the apparent objectivity of individual researchers: we are often unable to detect our own faulty reasoning.

How does Longino's view on objectivity concern individual COIs? The criteria for objective communities are not designed for the evaluation of individuals, and thus do not directly ban individual COIs. Like cultural or political interests, financial COIs can be revealed in discussions within and between communities. Indeed, one possible way of responding to the worry over the biasing influence of COIs is to argue that they do not impede objectivity but actually promote it by increasing the diversity of viewpoints. An argument like this has been advocated by Carrier (2010), who states that the competition between companies funding studies provides reasons for the sponsored scientists to examine the work of their competitors in more detail: researchers' COIs create the pluralism needed for cultivating criticism when parties try to beat each other in the race for creating the most lucrative product. However, even if a group of researchers had ties to competing agents, their background assumptions concerning the object of study and the interests steering their work may be in line, which, in turn, can guide their research towards similar outcomes.[5]

---

[5]I do not claim that financial interests determine the outcomes of studies. However, as I shall state below, there is statistical evidence on the steering effect the source of funding has on research

COIs constitute a threat to objectivity if they steer research unduly and, for one reason or another, the social critical mechanism revealing their influence fails to function properly. If the conflict concerns only financial interests, or if it is in the interest of all parties not to question certain assumptions, there is no reason to suspect that financial ties would promote critical discussions. For Longino, the diversity that counts is the diversity of background assumptions that influence our reasoning, and researchers can have conflicting conflicts of interests without conflicting background assumptions. If financial ties that researchers have work against forming diverse approaches to the object of the study, the diversity supporting objectivity can be said to be threatened. Carrier does notice that in some fields diversity does not follow commercialization. If the interests of the sponsors are the same, and serve as a motive to let some of the claims made by competitors stay uncontested, long standing biases may be left unquestioned. The case Carrier (2010) mentions is the research on the link between smoking and cancer. In what follows, I shall give another example of a situation where the social mechanism failed to function.

It is well known that in medical studies there is a statistical correlation between the source of the funding and the results: industry-sponsored studies reach positive results more often than studies funded by other agents, and studies funded by companies also tend to favor new products more often than studies without industry-funding—even if the new product tested was produced by a competing firm (e.g., Bekelman, Li and Gross 2003; Stelfox et al. 1998; Lundh et al. 2012). Despite the fierce competition and amounts of money involved, there are issues with respect to which the funders have shared interests. One example is the possible link between SSRIs (Selective serotonin reuptake inhibitors), used for treating depression and multiple other ailments, and suicidal tendencies. Several pharmaceutical companies have developed their own brands of SSRI, for instance GlaxoSmithKline (former SmithKline Beecham) has Paxil, Pfizer Zoloft and Eli Lilly Prozac. Thus, these competitors have been conducting research on products of similar type.

According to Healy (2002a), the first articles suggesting a connection between SSRIs and suicide were published in 1990. Later during the decade, evidence on the suicidality being a class effect started to accumulate. However, this was dismissed because of the lack of specifically designed trials to detect the effect.[6] If meta-analyses of all the data had been executed, the risk of suicide attempts would have been detected earlier (Healy 2006; Healy and Whitaker 2003). This was confirmed in 2005 when Fergusson et al. published a study where they systematically reviewed all the published randomized controlled trials (RCTs) to examine the

---

(Footnote 5 continued)
outcomes. In nutrition research (Lesser et al. 2007), research on tobacco smoking (Bero 2005), and medical research (Lundh et al. 2012), for example, there seems to exist a link between the source of funding and the outcomes of studies. Thus, it is pertinent to examine this issue further.

[6]The official black box warning announcing that SSRIs may cause suicidality was attached to packages no earlier than in 2004.

alleged link between the use of SSRIs and suicide attempts. The result was that there was "a more than twofold increase in the rate of suicide attempts in patients receiving SSRIs compared with placebo or therapeutic interventions other than tricyclic antidepressants" (Fergusson et al. 2005, 398).

Further, Healy argues that in addition to the lack of trials looking particularly into the possible connection between the drug and self-destructive behavior, the trials on SSRIs suffer from methodological limitations and flaws. According to him, many published trials were ghostwritten, did not report suicidal acts, reported that the acts had happened in the placebo group instead of the group taking the active treatment, or listed the acts in a misleading way (Healy 2006, 2011). Fergusson et al. support Healy's worries: "A number of major methodological limitations of the published trials may have led to an underestimate of the risk of suicide attempts" (Fergusson et al. 2005, 402). Study by Whittington et al. (2004) backs Healy's claim about parts of the data being left out of the analyses: their meta-analysis reveals that published and unpublished data paint very different pictures on the efficacy and safety of treating children with SSRIs.

Ghostwriting and not publishing negative studies are not the only suspicious practices related to medical research, and Healy is only one of the authors to draw attention to the problematic practices in this field of study and particularly in randomized controlled trials. Even though the RCTs are seen as a very reliable method of providing evidence, the design of the experiment and a fitted interpretation of the data can be used to tailor the results[7]: Smith (2005), the former editor of British Medical Journal, has claimed that firms use dubious methods, such as conducting trials of their own drugs against treatments known to be inferior or selecting only parts of the data for analysis, in order to achieve the wanted results. It has also been suggested that by choosing the subjects of the studies in a certain way, it is possible to influence the outcomes of the studies (Brown 2010; Petryna 2007). Bekelman et al. (2003) found out that using inactive control, such as placebo, was more common in studies funded by the industry than in studies funded by other sources, and that using placebo increased the likelihood of positive results.

Did the research on SSRIs fail to be objective? In other words, were claims on the drugs' efficacy and safety evaluated in a manner that reflected the preferences of involved parties? The above mentioned reports on trials suggest that research was designed to reach outcomes that concealed the adversary effects of the drugs: for example, the initial studies were too small to detect suicidality, suicidal behavior was not reported properly, and parts of data were left out of the analyses. In addition, the publicity of research, an essential condition for objectivity, was not realized, as data on adversary effects was not published. As a result, SSRIs appeared to be safer than subsequent research has revealed them to be. However, this does

---

[7]The discussion of the role of RCTs and how they ought to be monitored is connected to the debate over FDA's relationship with the pharmaceutical industry and how the system through which drugs are approved has motivated the industry to adopt dubious lines of conduct. Because of the lack of space, this matter cannot be examined here. These questions have been discussed, e.g., by Biddle (2007) and Healy (2012).

not yet entitle the conclusion that financial interests were responsible for the bias. To establish this, data on the intrusion of extraneous factors would be needed. And indeed, evidence on the influence of commercial interests is available. As mentioned above, the source of the funding correlates with the results so that industry-sponsored studies tend to reach pro-industry conclusions. Surely, correlation does not imply causation. However, empirical work on pharmaceutical research can reveal mechanisms through which the outcomes of studies are tinkered: Sergio Sismondo's (2007, 2009) work on the so-called ghost management of scientific research has helped shed some light on the practices the industry has adopted to reach favorable conclusions in pharmaceutical research. When research is ghost managed, it is designed, conducted and finally reported according to the plans of the industry. In addition, Healy and Cattell (2003) compared industry-linked articles on SSRI to non-industry-linked articles.[8] The articles linked to industry seemed to be ghost managed. Further, they had all reached positive results and underreported side effects. All in all, data indicates that the results of the research on SSRIs have indeed reflected the preferences of the industry.

However, the possible side effects of SSRIs were eventually officially recognized. Who then was the one to call attention to the inadequacy of the studies on treating children with SSRIs? According to Healy (2002b), journalists and lawyers were the first to make critical questions about the details of the trials—not the scientists. In other words, the critical voices came from outside the community of researchers. What sort of a lesson does this case teach us with respect to Longino's view on objectivity? First of all, the case of SSRIs demonstrates how certain ways of conducting research may not be criticized if basically everyone in the research community has similar interests: the researchers' possibility to question certain assumptions was limited because of their financial ties to the actors that, despite being competitors, had shared interests. In addition, the withholding of relevant data slowed down the critical interactions. This example shows how financial interests do not always help to add more competing points of view to a field of study but, on the contrary, contribute to the development of a situation where assumptions that steer research and entitle certain choices are not questioned. Healy's diagnosis of the case involving SSRIs and suicidal behavior is that the individual biases shared by practically all researchers caused the possible danger of the products to go unnoticed:

> If the marketplace worked properly and brought competing compounds into the therapeutic arena at the same time, we might be able to depend on companies to ferret out the hazards of their competitors' compounds. But in practice, possibly because of current patenting arrangements, new agents come to the market in classes, and this means that none of the companies sponsoring any of these agents has any incentive to detect what may be class-based problems. (Healy 2002a, 259)

If Healy's interpretation is correct, epistemic endeavors were dependent on institutional arrangements that are generally considered to be innocuous with

---

[8]Information on the industry articles had become available through legal action.

respect to science. The fact that information on the tactics of pharmaceutical industry has become public through legal actions adds to the point (e.g., Healy and Cattell 2003). This buttresses Longino's stance on the importance of paying attention to the institutional context of research.

4.2 Research Funding and Agenda Setting

As I shall argue later, the complications caused by COIs discussed above are partly related to the funding structure of the field. Funders, both private and public, naturally have their own motives and interests for sponsoring research and choosing the projects to support. This, consequently, shapes research agendas. Next I will explore this issue from the point of view of Longino's theory.

Funders do not consider all fields of study or approaches to be as attractive as others, an issue discussed by authors such as Brown (2010), Irzik (2007) and Kitcher (2001, 2011). Here I shall study the topic by presenting an example on how research funding and the development of a dominant approach have been intertwined in psychiatry. Pharmaceutical companies fund approximately 60 % of medical research (Musschenga et al. 2010). In a study conducted by Dorsey et al. (2009), the industry was found to be the biggest sponsor of biomedical research across all therapeutic areas, excluding the research on HIV/AIDS, infectious disease research and oncology. Therefore the field is highly influenced by the decisions made by industry.

Wyatt and Midkiff (2006) and Musschenga et al. (2010) discuss what kinds of explanations are searched for when the causes and cures for mental illnesses are studied. The argument in both papers is that in psychiatry important areas are left unexamined and questions unasked since the dominant approach steers researchers to heed only certain aspects of the investigated phenomena: Musschenga et al. (2010, 122) call it "the medical model", which "does privilege biology over other disciplines". Wyatt and Midkiff (2006, 132) use the term "biological psychiatry": "It reflects growing acceptance of the notion that chemical imbalances, genetic defects and related biological phenomena cause disorders such as schizophrenia, depression, anxiety, substance abuse, and attention deficit hyperactive disorder (ADHD)". The authors argue that this view is weakly supported by evidence but has gained popularity both among the general public and professionals (Wyatt and Midkiff 2006). Neither Wyatt and Midkiff nor Musschenga, Van der Steen and Ho want to argue that there are no disorders caused by genetic defects or other biological factors. What they state is that the current dominance of "biological psychiatry" or "medical model" is unwarranted and partly maintained by the funding structure of medical research.

It is understandable that the pharmaceutical industry is interested in funding research that may result in patentable products. As Wyatt and Midkiff (2006, 134) remark, "biological causation suggests biological treatment, rather than behavioral intervention". Biological treatments, i.e. drugs, are patentable while diets and behavioral therapies are not, and thus, from the industry's point of view, it is profitable to invest in research projects that search for the biological causes of mental disorders. Consequently, the resources for alternative approaches are scarce, and for researchers competing for funding it is safer to adopt a view that is most likely to attract sponsoring. This, in turn, hinders the development of approaches questioning the assumptions behind the dominant biological model.

Because of the ascendancy of the approach concentrating on the biological causes of mental illnesses, the research on diet, infections, biological rhythm and other environmental factors possibly influencing mental health has been unjustly disregarded (Musschenga et al. 2010; Wyatt and Midkiff 2006). This has led to problems of both theoretical and practical nature: For example, if circadian rhythms have an impact on the effectiveness of medication and this is disregarded in randomized controlled trials, the reliability of these trials is threatened (Musschenga et al. 2010). Adopting the view that one's problems are caused by one's biological make-up may help to ease the possible guilt often related to mental illnesses, but it can also make it less likely for one to seek help from therapy or to try to improve one's coping skills (Wyatt and Midkiff 2006).

What kinds of problems concerning objectivity arise if the above described example holds true? Remember that according to Longino, reaching objectivity is dependent on the availability of critical points of view. If alternative voices do not have the resources to develop, there is a danger that the values and assumptions that may steer research towards certain research topics, questions and approaches are not questioned. Consequently, the example on how the sources of funding have influenced the research agenda in psychiatry serves as a demonstration of a situation where the conditions for objectivity are potentially threatened by contextual factors. It helps to elicit the kinds of factors which can impact the direction of research, and thus, guide philosophers of science towards the material that can be utilized for improving theoretical development. Empirical work on the allocation of funding could offer valuable resources to philosophers interested in advocating diversity. For example, Lamont (2010) has studied how theoretical pluralism can be sustained in funding decisions.

5 Remarks on the Cases

Above I have presented two examples of research influenced by the context it is conducted in. Via these cases I have intended to clarify some of the problematic issues that philosophical theories of current science need to accommodate: How can the available sources of funding contribute to one perspective gaining dominance within a field of study? In what situations may extra-scientific interest threaten objectivity? By adopting Helen Longino's definition of objectivity, it is possible to identify in what way the research on SSRIs failed to be objective and what it actually is that makes the dominance of "biological psychiatry" seem dubious. In particular, the discussion has aimed at demonstrating how epistemic problems can have non-epistemic roots, which may be explored using empirical and more purely philosophical means.

Slaughter and Leslie (1999) argue that in response to increased global competition, countries have developed their R&D and higher education policies, which in turn is changing the working conditions within academia and other research organizations: when state funding for research becomes scarcer, institutions have to start competing for external funding and adapt their organizations and functions to best attract resources. It seems that increasing competition for private funding has forced research communities to adopt some tactics that make the development of critical points of view more difficult. The epistemic malfunctions resulting from one-sided funding are particularly problematic in fields the results of which bear direct consequences for the well-being of the public. For example, even though at the individual level the risk of suicide attempts related to using SSRIs remains relatively low, it should be considered to be a serious threat at the population level due to the extensive consumption of the drugs (Fergusson et al. 2005).

What, then, could be done about the one-sided funding of certain fields? Carrier (2010), Musschenga et al. (2010) as well as Schafer (2004) have a common suggestion for dealing with the problem: increasing the public funding of research to promote competing approaches and to advance diversity. However, as already mentioned, decisions on research funding are made outside the research community. This suggests that the conditions for achieving objectivity, understood in Longino's sense, are not dependent only on the actions of the researchers. The diversity of opinions within research communities and the possibility of inter-communal discussions are dependent on extra-scientific factors, such as the science and higher education policy—conducted both on national and international levels—that shapes the functions of communities.

In addition to drawing attention to the significance of contextual factors, the examined cases also offer support for Longino's social view on objectivity: Studying how commercial interests can steer research via funding can help us see how the integrity of individuals and rigorous testing of hypotheses are not enough for achieving reliable knowledge if the community-level mechanisms are not functioning properly. As Brown (2010) has argued, the so-called "discovery side of science", i.e., the generation of hypotheses and research questions, is not innocuous with respect to objectivity. If a mechanism somehow prevents the development of alternative explanations for a phenomenon, e.g., a mental disorder, it is more difficult to critically inspect the assumptions steering research. In other words, even if one is not willing to agree with Longino in that the interested reasoning of an individual does not have to hinder objectivity, one should nonetheless take seriously her request to paying attention to the context of research activities.

The examples of SSRIs and so-called biological psychiatry demonstrate how philosophy of science can benefit from empirical work. For instance, Sergio Sismondo's research on the so-called ghost management of medical studies can be used to explicate how new actors, such as publication planners, enter the arena of research, how this affects the processes that are designed to secure objectivity, and how, consequently, normative theories should be modified to better accommodate new ways of practicing research. In a similar vein, studies on research funding (e.g., Lamont 2010; Slaughter and Leslie 1999) illuminate what kinds of factors are involved in the allocation of resources for different approaches. These studies offer insight into an important condition for the plurality of points of view, and thus, for objectivity.

References

Babcock, L., Loewenstein, G., Isscharoff, S., Camerer, C.: Biased judgment of fairness in bargaining. Am. Econ. Rev. **85**(5), 1337–1343 (1995)
Barwell, I.: Towards a defense of objectivity. In: Lennin, K., Whitford, M. (eds.) Knowing the Difference. Feminist Perspectives in Epistemology, pp. 79–94. Routledge, London (1994)
Bekelman, J., Li, J., Gross, C.: Scope and impact of financial conflicts of interest in biomedical research. JAMA **289**(4), 454–465 (2003)
Bero, L.: Tobacco industry manipulation of research. Public Health Chronicles. **120**(2), 202–208 (2005)
Biddle, J.: Lessons from the Vioxx debacle: what the privatization of science can teach us about social epistemology. Soc. Epistemol. **21**(1), 21–39 (2007)
Brown, J.R.: One-Shot Science. In: Radder, H. (ed.) The Commercialization of Academic Research. Science and the Modern University, pp. 90–109. Pittsburgh University Press, Pittsburgh (2010)
Carrier, M.: Research under pressure. Methodological features of commercialized science. In: Radder, H. (ed.) The Commodification of Academic Science. Science and the Modern University, pp. 158–186. Pittsburgh University Press, Pittsburgh (2010)
Dorsey, E.R., Thompson, J.P., Carrasco, M., de Roulet, J., Vitticore, P., et al.: Financing of U.S biomedical research and new drug approvals across therapeutic areas. PLoS ONE **4**(9), e7015 (2009). doi:10.1371/journal.pone.0007015
Fergusson, D., Doucette, S., Cranley, Glass, K., Shapiro, S., Healy, D., Hebert, P., Hutton, B.: Association between suicide attempts and selective serotonin reuptake inhibitors: systematic review of randomised controlled trials. BMJ **330**, 396–402 (2005)
Healy, D.: Conflicting interests in Toronto. Anatomy of a controversy at the interface of academia and industry. Perspect. Biol. Med. **45**(2), 250–263 (2002a)
Healy, D.: In the grip of the python: conflicts at the university-industry interface. Sci. Eng. Ethics **9**, 59–71 (2002b)
Healy, D.: The antidepressant tale: figures signifying nothing. Adv. Psychiatr. Treat. **12**, 320–328 (2006)
Healy, D.: Science, rhetoric and the causality of adverse events. Int. J. Risk Saf. Med. **24**, 1–14 (2011)
Healy, D.: Pharmageddon. University of California Press, Berkeley (2012)
Healy, D., Cattell, D.: Interface between authorship, industry and science in the domain of therapeutics. Br. J. Psychiatry **183**, 22–27 (2003)
Healy, D., Whitaker, C.: Antidepressants and suicide: risk-benefit conundrums. J. Psychiatry Neurosci. **28**, 331–337 (2003)
Irzik, G.: Commercialization of science in a neoliberal world. In: Bugra, A., Agartan, K. (eds.) Reading Polanyi for the 21st Century: Market Economy as a Political Project, pp. 135–153. Palgrave Macmillan, New York (2007)
Irzik, G.: Why should philosophers of science pay attention to commercialization of academic science? In: Suárez, M., Dorato, M., Rédei, M. (eds.) EPSA Epistemology and Methodology of Science: Launch of the European Philosophy of Science Association, pp. 129–138 (2010)
Jukola, S.: The commercialization of research and the quest for the objectivity of science. Found. Sci. (2014). doi:10.1007/s10699-014-9377-9

Katz, D., Caplan, A., Merz, J.: All gifts large and small: toward an understanding of the ethics of pharmaceutical industry gift-giving. Am. J. Bioeth. **3**(3), 39–46 (2003)

Kitcher, P.: Truth, and Democracy. Oxford University Press, Oxford (2001)

Kitcher, P.: Science in a Democratic Society. Prometheus Books, New York (2011)

Krimsky, S.: Science in the Private Interest. Has the Lure of Profits Corrupted Biomedical Research? Rowman and Littlefield Publishers, Inc., Lanham (2003)

Lamont, M.: How Professors Think? Inside the Curious World of Academic Judgment. Harvard University Press, Cambridge (2010)

Lesser, L., Ebbeling, C., Goozner, M., Wypij, D., Ludwig, D.: Relationship between funding source and conclusions among nutrition-related scientific articles. PLoS Med. **4**(1), e5. doi:10.1371/journal.pmed.0090005 (2007)

Leuschner, A.: Pluralism and objectivity: exposing and breaking a circle. Stud. Hist. Philos. Sci. **43**, 191–198 (2012)

Longino, H.: Science as Social Knowledge. Princeton University Press, Princeton (1990)

Longino, H.: Gender, politics, and theoretical virtues. Synthese **104**, 383–397 (1995)

Longino, H.: The Fate of Knowledge. Princeton University Press, Princeton (2002)

Lundh, A., Sismondo, S., Lexchin, J., Busuioc, O.A., Bero, L.: Industry sponsorship and research outcome. Cochrane Database Syst. Rev. (Issue 12). Art. No.: MR000033 (2012). doi:10.1002/14651858.MR000033.pub2

Musschenga, A., Van der Steen, W., Ho, V.: The business of drug research: a mixed blessing. In: Radder, H. (ed.) The Commodification of Academic Science. Science and the Modern University, pp. 110–131. Pittsburgh University Press, Pittsburgh (2010)

Petryna, A.: Clinical trials offshored: on private sector science and public health. Bio Soc. **2**, 21–40 (2007)

Schafer, A.: Biomedical conflicts of interest: a defense of the sequestration thesis—learning from the cases of Nancy Olivieri and David Healy. J. Med. Ethics **30**, 8–24 (2004)

Shamoo, A., Resnik, D.: Responsible Conduct of Research, 2nd edn. Oxford University Press, Oxford (2009)

Shapin, S.: Scientific Life. University of Chicago Press, Chicago (2008)

Sismondo, S.: Ghost management: how much of the medical literature is shaped behind the scenes by the pharmaceutical industry? PLoS Med **4**(9): e286. doi:10.1371/journal.pmed.0040286 (2007)

Sismondo, S.: Ghosts in the machine. publication planning in the medical sciences. Soc. Stud. Sci. **39**(2), 171–198 (2009)

Slaughter, S., Leslie, L.: Academic Capitalism. Politics, Science, and the Entrepreneurial University. The John Hopkins University Press, Baltimore (1999)

Smith, R.: Medical journals are an extension of the marketing arm of pharmaceutical companies. PloS Med. **2**(5), e138 (2005)

Smith, T.: "Social" objectivity and the objectivity of values. In: Machamer, P., Wolters, G. (eds.) Science, Values, and Objectivity, pp. 143–171. Pittsburgh University Press, Pittsburgh (2004)

Stelfox, H., Chua, G., O'Rourke, K., Detsky, A.: Conflict of interest in the debate over calcium-channel antagonists. N. Engl. J. Med. **338**, 101–106 (1998)

Whittington, C.J., Kendall, T., Fonagy, P., Cottrell, D., Cotgrove, A., Boddington, E.: Selective serotonin reuptake inhibitors in childhood depression: systematic review on published and unpublished data. Lancet **363**, 1341–1345 (2004)

Wilholt, T.: Bias and values in scientific research. Stud. Hist. Philos. Sci. **40**, 92–101 (2009)

Wyatt, W.J., Midkiff, D.M.: Biological psychiatry: a practice in search of a science. Behav. Soc. Issues **15**, 132–151 (2006)

Part III
Empirical Philosophy of Science and HPS

History and Philosophy of Science as an Interdisciplinary Field of Problem Transfers

Henrik Thorén

Abstract The extensive discussions of the relationship between the history of science and the philosophy of science in the mid-20th century provide a long history of grappling with the relevance of empirical research on the practices of science to the philosophical analysis of science. Further, those discussions also touched upon the issue of importing empirical methods into the philosophy of science through the creation of an interdisciplinary field, namely, the history and philosophy of science. In this paper we return to Giere (1973) and his claim that history of science as a discipline cannot contribute to philosophy of science by providing, partial or whole, *solutions* to philosophical problems. Does this imply that there can be no genuine interdisciplinarity between the two disciplines? In answering this question it is first suggested that connections between disciplines can be formed around the transfer and sharing of *problems* (as well as solutions); and that this is a viable alternative for how to understand the relationship between history and philosophy of science. Next we argue that this alternative is sufficient for establishing a genuine form of interdisciplinarity between them. An example is presented—Darden's (1991) book on theory change—that shows how philosophy of science can rely on history of science in this way.

Keywords History and philosophy of science · Interdisciplinarity · Problem transfer · Problem feeding

1 Giere's Characterization of HPS

Current debates on how philosophy of science can be informed by ethnographic and sociological case studies run parallel to debates from the 1970s and 1980s on how philosophy of science can and cannot be informed by historical case studies.

H. Thorén (✉)
Department of Philosophy, Lund University, Lund, Sweden
e-mail: Henrik.thoren@fil.lu.se

Investigating this parallel we depart in this paper from Ronald Giere's widely disseminated marriage of convenience metaphor for the relationship between history and philosophy of science. The metaphor—popular since its conception—was proposed by Giere in his (1973) paper "History and Philosophy of Science: Intimate Relationship or Marriage of Convenience?". The paper reviewed the contents of the fifth volume of *Minnesota Studies in the Philosophy of Science* on "Historical and Philosophical Perspectives on Science". Giere complained, that among the papers in that volume he found only pure history papers, or pure philosophy of science papers. This observation prompted him to ask the following question:

> Now let us grant that philosophy of science without science would be empty. The question for one holding the "Kantian" dictum is whether and how the historian of science, as historian, has anything essential to contribute to the content of contemporary philosophy of science. (Giere 1973, 286)

The question was, in part, motivated by the numerous departments, centres, and programmes devoted to the history and philosophy of science (henceforth HPS) that had become fairly common around that time, at least in the US. This development may have been taken as an indication of an increase in the intellectual exchange between the disciplines but Giere remained sceptical. These new departments and centres might just as well be a common refuge for two sub-disciplines trying to slip the confines of their parental homes. All it really "shows [is] that neither historians nor philosophers of science are happy with their parent disciplines" (Giere 1973, 296). Hence the marriage of convenience metaphor. Giere has later recalled that the department at which he himself was active at the time—the Department of History and Philosophy of Science at Indiana University—was, quite in spite of its name, not a place where a great deal of integration or communication was going on between the two disciplines. The separation was even manifested physically as "all the historians' offices were on one side of the hall and the philosophers' offices on the other" (Giere 2011, 59).

Another reason the relationship between history and philosophy of science was of interest at this time was that it acted as one battleground in the larger debate concerning the axiomatic conception of science (Schickore 2011, 456). Some philosophers were—in contrast to earlier positivist ideas of an ahistorical philosophy of science—arguing that philosophy of science should be, or by necessity was, "inextricably intertwined" with the history of science (ibid.). It was against such allusions to interdisciplinary intimacy that Giere voiced his doubts, arguing not only that the relationship was indeed a marriage of convenience, but also that it could be nothing else. Philosophy of science *qua* philosophy cannot draw on history of science *qua* history.

For later comparisons I will first translate Giere's claim about HPS into the language of interdisciplinarity. From this perspective, what he seems to claim is that there is no genuine interdisciplinarity between the disciplines of history of science and philosophy of science.

2 The Descriptive and the Normative

The debate on the state of HPS of the 1960s and the 1970s—in the context of which Giere's contribution should be understood—concerned a number of different questions. Some concerned how historical information could influence normative philosophical analysis, others, for example, how to do proper history, and how to characterize philosophical analysis itself (Schickore 2011, 455). In introducing the marriage metaphor, Giere was concerned mainly with the first of these questions. His main reason for being sceptical about the contributions of history to the philosophy of science had to do with the is/ought-distinction; in the absence of an account of how to derive normative conclusions from descriptive statements no philosophical issues can be determined from historical facts:

> If one grants that epistemology is normative, it follows that one cannot get an epistemology out of the history of science—unless one provides a philosophical account which explains how norms are based on facts. (Giere 1973, 290)

In other words, a descriptive approach, such as history of science, can never inform a normative approach, such as philosophy of science.

However, the first quotation extracted from Giere starts by accepting Lakatos' (1971) observation that philosophy of science without science is empty. But if we accept Lakatos' observation, we should not be so quick to deny that history of science sometimes informs philosophy of science. The historian can tell the philosopher of science something about science. On the assumption that philosophy of science without science is empty, then clearly what the historian knows can sometimes be sufficient to further philosophy of science. Exactly how and why partly depends on the way in which philosophy of science without science is empty. I will return to this question below. Suffice it to say here that it is even likely that the historian—being interested in descriptive matters—rather than the philosopher of science—with her interest in normative matters—has access to facts about science. In other words, accepting Lakatos' observation, that there should never be any intimate relations between history and philosophy.

Hence, another and more fruitful interpretation of the claim that one cannot get an epistemology out of the history of science is that a descriptive approach, such as history of science, can never by itself solve certain kinds of problems a normative approach, such as philosophy of science, has identified.

The difference between the two interpretations can be pictured by deploying the traditional distinction between the context of discovery and the context of justification: claiming that a descriptive approach can never inform a normative approach denies that history has a role to play in either of the two contexts, while claiming that a descriptive approach can never by itself solve certain kinds of problems arising within a normative approach denies that history has a (considerable) role to play in the context of justification.

Finally, Giere also adheres to the claim that, in any case, it is not necessary that philosophy of science is informed by history of science. According to Lakatos' observation, what philosophy of science needs, is to get in touch with science. However, there appear to be a number of ways this can be achieved without involving history of science. One can get historical facts about science from elsewhere; from science itself, for example. A second possibility is to access non-historical scientific facts. Giere argues that this would be even better than accessing historical scientific facts:

> Philosophers and scientists may be influenced by their understanding of historical cases. But history of science need not enter the process, and it would be difficult to argue that it should. What we seek is a unified method of validation to be applied in current scientific inquiry. To argue that our understanding of past science, which is itself based on empirical evidence, should be fed in the process of choosing a theory of validation is to assume that we are right about the past and that this past experience is relevant to present scientific inquiry. (Giere 1973, 294)

Hence, on Giere's view in the 1973 paper, history of science can by itself never solve problems arising within philosophy of science, and nor is it necessary that philosophy of science is informed by history of science.

Giere's sceptical conclusion is based on particular ideas about the nature of these respective disciplines. Philosophy is conceived of as dealing with normative issues of science whereas history is confined to the descriptive. Although there is a considerable literature that questions just exactly how "pure" is the context of justification—that a normative philosophy of science would be confined to—the standard challenge to Giere's sceptical remarks involves adopting a different idea of what philosophy of science is.

In later writings, Giere has changed his mind on the nature of philosophy of science (Giere 1988, 2011). The naturalized philosophy of science encompassed in his cognitive approach has other goals than mere prescription—it aims to "construct a theory of how science works" (Giere 2011, 61). This project is deeply empirical and draws on a number of other disciplines; cognitive science, sociology of science, and anthropology of science, for example, history of science is among these but has no privileged role.

A second influential example is Laudan (1989) who thought of philosophy of science as the project of establishing theories of theory change and envisaged history as providing data for philosophy of science, against which these theories could be tested. As the preferred form of this data was longitudinal accounts of theory change Laudan's conception clearly gives history of science a special, and unique, position with respect to philosophy of science.

Finally, a third conception, recently defended by Schickore (2011), puts the emphasis on understanding. Schikore argues that philosophical analysis leans more towards hermeneutics than a science of science (as Laudan and Giere imagine). On this model, history is built into the very core of the philosophical project; a crucial part of knowing what science *is* and what makes it productive, simply is, to know how it came about.

3 More Than a Marriage of Convenience

These later developments aside, even if we accept the position that history of science can never by itself solve problems arising within philosophy of science, nor is it necessary that philosophy of science is informed by history of science, it is still misleading to think of the relationship between history of science and philosophy of science as a marriage of convenience. Instead, it seems to characterise most marriages between disciplines that the one can never by itself solve the problems arising within the other, nor is it necessary that the one discipline informs the other discipline, History of science and philosophy of science manifest *some* kind of genuine interdisciplinarity. But what kind?

The idea pursued in this article is that the generation and transfer of problems is a genuine interdisciplinary activity—including the generation and transfer of problems between history and philosophy of science. That interdisciplinarity can be conceived as the transfer of elements between two disciplines, is not a new idea. For instance, Mitchell et al. (1997) discusses a number of transfers (of tools, metaphors, models, and techniques) that they think answer 'the whys and hows of interdisciplinarity.'[1] This will be returned to in Sects. 4 and 5. Here the argument is that the transfer of problems could potentially have a fundamental place in such an account of interdisciplinarity in history and philosophy of science.[2]

This task will be approached by first by pointing to the centrality of problems within disciplines. and then to the fact that problems are sometimes transferred between disciplines. In the next section, an example is offered that highlights the purpose of problem-transfers between history of science and philosophy of science.

The relationship between disciplines and problems is multifaceted. First, problems are sometimes thought to be the very locus of disciplines; that is to say, particular disciplines define their domain of inquiry by reference to a set of problems. When Darden and Maull (1977) in their influential paper on interfield theories developed their notion of a scientific field—which they themselves thought to be a roughly similar to a discipline[3]—a central problem is by far the most important component. Second, disciplines are also the source of new problems. Generally speaking, problems arise out of specific theoretical contexts upon which they depend (Nickles 1981; Laudan 1977; Toulmin 1972). The tension between, on the one hand theories, expectations, explanatory ideals, and so on, and on the other perceived states of affairs (observations, for instance) is what generates new problems. Disciplines are the contexts which provide all of these components. And

---

[1] See also Thorén and Persson (2011).

[2] Transfers of problems between disciplines is likely to often involve some type of transformations. Furthermore, as Grantham (2004) has pointed out sometimes one discipline use another as a resource of interesting problems and hypotheses. See Thorén and Persson (2013).

[3] They compare fields with Toulmin's conception of a discipline and deem them to be more or less the same, although they prefer their own terminology as to avoid confusion with Toulmin's approach to science. See Darden and Maull (1977, 45).

third, disciplines have, by tradition, access to (or expertise in) particular methods, tools, and approaches that make them more or less suitable to solve particular problems.

The connection between coming across a problem and being able to solve it is less than rigid; a discipline may discover a problem that cannot be solved within that discipline (given how it is constituted at the time of discovery). This sometimes leads to interdisciplinarity, as has been recognized within the literature on interdisciplinarity for quite a while. Sherif and Sherif (1969), for example, consider in brief the case of metabolic researcher Dr. William Schottstaedt. Schottstaedt, while conducting a study in his metabolic ward, discovered that interpersonal relationships apparently had an influence on metabolic measures. In order to explain the measures he obtained, he would have to venture well beyond his disciplinary expertise. Perhaps a more suitable approach at this point might be to engage in a sociological inquiry? Sherif and Sherif do not disclose how the case developed but two possibilities appear to have confronted Schottstaedt; either he *export* the problems or he *import* the necessary cognitive resources. There is probably no general guidance as to what is the best line of action but a lesson that can be drawn from this; that a problem is generated, or discovered, within a particular discipline does not entail that the problem will be possible to solve within that discipline.[4]

Others too have noted on similar kinds of problem transfers. Maull (1977) discuss problems that shift between appropriately related fields. These problems are preceded by shared terminology and find their solution in interfield theories.[5] One important kind of situation is what Grantham (2004) calls *heuristic dependence*. Certain fields,[6] or disciplines, may depend on others for formulating hypotheses. For example, neuroscience may look to psychology in order to obtain problem formulations and philosophers of biology might look to biology for theirs. Whereas Schottstaedt might have been prompted to export his problems to someone with the appropriate expertise, it is also clear that some disciplines draw on others for their problems. They *import* their problems, so to speak.

The transfer of problems in HPS has to do with heuristic dependence. Two qualifications to this observation are needed. First, one may argue that in the case of heuristic dependence, but not in the case of import and export of problems, the problem arises from the *interaction* between the disciplines. It is doubtful, however, that this constitutes a sharp distinction. It depends on how one determines to what extent observations 'belong' to a discipline or not. Second, whereas Schottstaedt apparently discovered that the problem *he was interested in solving* was not one

---

[4]Again, more could be said about this. A relevant fact here is that disciplines generally are not isolated contexts but occur in broader contexts and that the scientists active within a discipline will have perspectives that go beyond their working environment. Or so one would hope. These are facts that matter and make the placing of problems a little more difficult.

[5]See Darden and Maull (1977); also see Thorén and Persson (2013) for a discussion on problem transfers and interfield theories.

[6]Grantham uses the notion of a field that is due to Darden and Maull (1977). For our purposes we will take fields to be roughly the same as what we refer to a disciplines, see note 2.

that he could solve, given the present situation within his discipline, in many cases of heuristic dependence the problems extracted will not be considered to be interesting to the "source discipline" quite regardless of whether they can be solved there or not.

Both of these points relate to what it means for a particular problem to *belong* to a discipline. Consider the following: A problem can be said to *belong* to a discipline if:

A. it *arises* within that discipline, or;
B. the methods, tools, procedures, or explanatory models within that discipline are *appropriate* for solving the problem.

These two principles generate somewhat different outcomes; under A we should seek to acquire the appropriate resources (and thus expand our own discipline) and under B the problem should be out-sourced to wherever those resources are already available. Is a problem a philosophical problem *because* philosophers can solve it? Or is it a philosophical problem because it can be said to have arisen within the confines of philosophy-the-discipline? Under some conceptions of the nature of philosophy and history then the exclusion of history from solving philosophical problems is just trivial; should it ever be the case that history solves the "philosophical" problem, then the problem wasn't genuinely philosophical to begin with. But this requires a rigid conception of disciplines in general, or at least, these particular disciplines. One suspects it is never entirely clear when a discipline should appropriate a new methodology as opposed to outsourcing problems that are beyond the scope of the discipline at a certain point in time. Moreover there is always the risk or opportunity that new additions, perhaps even mere methodological ones, actually *change* the problem they were meant to solve.

Bilateral problem-feeding, or the exchange of problems and solutions to the benefit of both of the involved disciplines or fields admittedly requires a well-established, and moderately stable, relationship of mutual interest and trust (Thorén and Persson 2013). In what Schickore (2011) calls the confrontation model—exemplified by the later Giere's cognitive approach, or Laudan's theory testing idea—the two disciplines are thought of as involved in a relationship that approaches this ideal. Consider Laudan: theories of theory change were to be tested against the historical record and in order to do so, someone needed to provide such a history of science. This problem—that is, reconstructing history—would thus ideally be out-sourced to historians (cf. Schickore 2011, 464). Much to the disappointment of philosophers of science, historians were not particularly enthusiastic about the project and would not produce the kind of longitudinal studies of theory change that philosophers craved. Laudan thus concluded that philosophers would have to do their own history (Laudan 1989, 13).

Perhaps one might find reasons for disregarding the A-possibility above, along similar lines as philosophers have been disinterested in processes of discovery? How problems come to arise in, and in the A-sense belong to, a discipline is an unstructured process guided only by the whims of particular scientists. But then again, granted that problems cannot be abstracted away from their theoretical setting it seems strange to disregard precisely that setting. Looking at the historical record, from the point of view of philosophy, may produce problems and questions that are philosophical to the extent that they are appropriately solved by deploying philosophical tools, and methods and explanations. They will however not be 'philosophical' in that they may not connect to the *specific* problems that have traditionally been discussed. In this sense it may even seem plausible that certain problems would never have entered philosophy of science unless history of science had identified them.[7]

Lastly, there are in all probability cases where it is important to clarify precisely who identifies a particular problem and what happens in the transfer of problems, and then, solutions. Here both the agents and their values may come into play and be important in providing an analysis. At other times, it might be meaningful to disregard the finer grains; philosophers may come across, in the historical record, problems that they find interesting and are able to solve. Here we will think of such cases as transfers of sorts; they qualify, on this conception, as problem-feeding, albeit of a unilateral sort (see Thorén and Persson 2013).

4 An Example: Darden's Method

There is a trend within philosophy of science to deploy a methodology which leans heavily on case studies, drawn both from the historical record and the annals of contemporary science. We will now move to explore a particular such attempt, namely Darden's (1991) study of the developments within genetics and neighboring fields of the early 20th century and the research strategies deployed during this period. This study serves as a prime example of a kind of mixed approach to the study of science that employs both historical and philosophical analyses. No particular claims about the success or failure of her project at large will be made, a project to which I am sympathetic. The aim is rather to discuss the methodology underpinning it.

Darden's method, in short, involves close readings of published papers by biologists of the time. On the basis of this she makes rational reconstructions—idealized discovery strategies—that if they had in fact been deployed could have generated the actual results.

> My account lays out actual historical changes. The aim of the philosophical analysis is then to find general strategies, which I claim are "exemplified" in such historical changes. The strategies are my own proposals of methods that could have produced those changes. (Darden 1991, 5)

---

[7]The necessity claim here is in another sense perhaps too strong; a similar, but nonetheless different, context could of course also generate a specific problem. This is probably true of any problem. Nonetheless, it is *actually* the case that history of science does provide philosophy of science with this particular service.

There are two concerns—both of which Darden are well aware—that can be raised about her approach. One concerns the historical/descriptive part, the other the philosophical/normative.

From a historical point of view the approach has some well known weaknesses. The most important one has to do with historical accuracy. There is no guarantee that the processes which Darden describes are the ones that were actually used—in fact it is quite probable that they were not. In order to determine what discovery strategies were actually deployed, published material is a poor source as it is generally contrived *ex post facto* and is guided by various other motives beyond accuracy; vanity, bad memory, the style of journals, and pedagogical considerations all play a part. To even approach historical accuracy further sources would have to be recruited: notebooks, diaries, correspondence, interviews, etc. Darden readily admits this problem and circumvents it with ingenious simplicity, by abandoning the ideal all together. Her reconstructions are supposed to mirror rational strategies that *would have* resulted in the discoveries in question.

The other issue concerns the philosophical content, and reverses the issue. Even if Darden has no ambition to produce a historical reconstructing of these episodes of scientific discovery, she carries out her philosophy of science in very close proximity to these episodes, which she examines with exemplary thoroughness. She calls them "cases" and has a chapter towards the end titled "Summary of strategies from the historical cases" (Darden 1991, 226). It is probably fair to say that Darden is involved in case work. Now, case studies are riddled with problems (cf. Schickore 2011, 468). Can they do philosophical work at all? How is one to construe one's cases to begin with without contaminating them? And, how is one to generalize from them? Darden frames her strategies in general terms and suggests they do generalize, at least to contexts that are "relevantly similar" (Darden 1991, 17). Taken at face value, that doesn't say much perhaps, but be that as it may. The point is, to the extent that Darden is doing case work, she is susceptible to problems associated with that practice.

If we consider Darden's approach in light of Giere's concerns it might appear as if Darden is put in a difficult predicament: By blending history of science and philosophy of science she could be seen as ending up with the worst of both worlds; no accurate descriptions and no useful prescriptions. However, this view would be mistaken. Instead, Darden's approach shows that history of science *enriches* the philosophy of science by supplying interesting problems that can be pursued. Darden's aim is to uncover strategies of discovery. The problem of developing possible strategies that can reproduce the results of early geneticists, is in a way a problem that can only arise at the intersection of history and philosophy of science. At the same time, it also involves an expansion of what philosophy of science is.

5 Interdisciplinarity as Transfer

That the transfer of cognitive contents is a form of interdisciplinarity has been recognized by many (Mitchell et al. 1997; Thompson Klein 1990; Kellert 2008; Mäki 2009). In many of these accounts the focus is on exporting, importing, or even imposing e.g. theories, models and methods. As pointed out in Sect. 3, however, there is also a literature on the transfer of problems (Sherif and Sherif 1969; Maull 1977; Thorén and Persson 2013). One question that may be raised at this juncture is how this transfer of problems relate to another central notion in the literature on interdisciplinarity, that of *integration.*

In Sect. 3 it was further noted that problem-feeding comes in different forms; sometimes it is unilateral, sometimes bilateral. Whereas unilateral problem-feeding requires comparatively little—there need not even be communication going on—bilateral problem-feedings is a rather more substantive process. It requires either that the standards relating to the evaluation of proposed solutions are shared, or, if standards are not shared, that a degree of trust is established (Thorén and Persson 2013, 347f). Furthermore a common interest must exist. Unilateral problem feeding—what Grantham calls heuristic dependence—requires none of these things to be in place but Grantham nonetheless considers it to be a form of practical unification (Grantham 2004, 143).

Consider the following argument. One sometimes senses an uncertainty in the literature with respect to the 'interdisciplinary outcomes.' Should we have—or is there already—a discipline *History and Philosophy of Science*? Or, is it preferable that the disciplines are kept apart, but *in touch,* so to speak? Wylie (1995) suggests that the appropriate approach to studying science is *interdisciplinary science studies,* which draws on philosophy, history, anthropology, ethnography, sociology, and so on. Science is a complex phenomenon that cannot be exhaustively described from a single perspective. This interdisciplinary science studies approach involves both independence—the different perspectives need to remain different, otherwise there is no inter-disciplinarity—and integration. Wylie notes that philosophy of science and sociology of science—for so long entangled in fierce dispute—now seem to have abandoned the battlements and started to approach one another. When Giere raised his concerns in 1973 it was in the context of a re-invigorated field that, at least on the surface, began to take the shape of a discipline (or sub-discipline). Giere, however, approached matters from the positivist conception of what philosophy of science is, or should do. Now, even this philosophy of science needs to stay in touch with the science it purports to study; otherwise it can hardly be called a philosophy *of science.* On the assumption that philosophy of science is a normative project and that such a project is cut-off from facts about science by the is/ought dichotomy this connection becomes admittedly limited. But at least one tie always remains, namely that philosophy of science needs science as a source of problems. These problems are not necessarily problems that scientists think they have, or are interested in, but nonetheless arise out of their practices. Issues of justification and discovery, what theories, models, and concepts are, all have sprung from science itself. A minimal form of problem-feeding thus arises; or rather, is the very prerequisite for there being a philosophy of science at all. History of science is not necessarily this source, although it is a natural one.

It would be difficult to make the case that unilateral problem-feeding *always* or *necessarily* leads to further integration but if we look at the development of philosophy of science since Giere's (1973) it is obvious that the discipline has become ever more inclusive; especially by becoming increasingly reliant on history of science, but also on other empirical approaches. With respect to history this reliance takes many forms and cannot be easily captured in programmatic statements on the nature of and relationship between the disciplines (cf. Arabatzis and Schickore 2012).

6 Concluding Remarks

Over the past five or six decades philosophy of science has gone through substantive changes and is now a sub-discipline that is broader than it once was (Arabatzis and Schickore 2012). Moreover the contexts distinction that played such a big part in separating philosophy of science from empirical approaches has also been successively hollowed out (Nickles 2006). However, even if we would still maintain that history of science can never in itself solve certain kinds of problems in philosophy, or is it necessary that philosophy of science is informed by history of science, there is still a way in which the philosophy of science can be dependent on history of science—namely as a source of problems.

So, if we return to Giere's question. How could history of science *qua* history contribute to contemporary philosophy of science? Based on the account of interdisciplinary transfer of problems I shall argue that one way in which history of science qua history can contribute to philosophy of science is by providing a backdrop against which new and interesting problems can arise. I think this relationship of problem-transfer can and has proven to be quite fruitful in staking out new domains of inquiry for philosophy. This kind of interdisciplinary relation differs markedly from what might be called a *programmatic* conception of the relationship between history and philosophy of science. On this programmatic approach the idea is to, in a systematic and forward-looking fashion spell out in what way, in this case, history-the-discipline may help to *solve* entrenched philosophical problems. Whether or not this is plausible, in general, or concerning specific problems is difficult to say. There are two interconnected points to make. The first one is that on the problem-feeding account, which is at least part of the truth, it is unlikely that a program could be formulated. None is needed, and it is easy to see how history of science will be highly fruitful for philosophy of science anyway. In a sense, thinking of the relationship here as an exchange of problems is putting things rather openly; it is consistent with many more specific ideas of how this relationship is to be spelled out. This might appear displeasing to some, precisely as a consequence of this lack of specificity. However, this may also be considered an asset. More specific accounts of how this relationship may take form depart from narrow, and hence contingent, conceptions of what the two disciplines are. Thus they are almost certain to fail over time. The second point then concerns interdisciplinarity; the suggestion here is that perhaps history and philosophy of science is best seen as a genuine interdiscipline that draws its strength from these tensions rather than be defeated by them. The difference is then perhaps that, as appears to be the case for Darden, the historical record suggests problems that are suitably solved by use of philosophical methods albeit these problems appear to remain outside of the 'mainstream' of philosophy of science. The counterparts of this marriage never become indiscernible from each other.

When Giere characterized philosophy of science—and history of science for that matter—in 1973 he adopted a much too constrictive conception of disciplines. Disciplines in general are dynamic and changeable. This has been a central theme in this paper. Another point has been that history of science can provide a fruitful resource for philosophy of science by providing a backdrop against which new problems can arise. This turns Giere's suggestion on its head; whereas he was thinking of history as supplying solutions I am here suggesting that it might instead provide the problems.

Initially these points may appear to be detached from one another but there is a sense in which they are not. Namely, even a one-sided reliance by one discipline on another for problems tends to affect the recipient. Darden's approach is a case in point; by taking on a particular historical period in science she found a problem suitable for philosophical analysis. But adopting this problem also, inadvertently, involves abandoning some of what philosophy of science might have been. Indeed the trend is for philosophy of science to take on an ever more empirical approach drawing on a range of other disciplines where history of science remains important, perhaps the most important.

References

Arabatzis, T., Schickore, J.: Ways of integrating history and philosophy of science. Perspect. Sci. **20**(4), 395–408 (2012)
Darden, L.: Theory change in science: strategies from Mendelian genetics. Oxford University Press, New York (1991)
Darden, L., Maull, N.: Interfield theories. Philos. Sci. **44**(1), 43–64 (1977)
Giere, R.: History and philosophy of science: intimate relationship or marriage of convenience? Br. J. Philos. Sci. **24**(3), 282–297 (1973)
Giere, R.: Explaining science: a cognitive approach. University of Chicago Press, Chigaco (1988)
Giere, R.N.: History and philosophy of science: thirty-five years later. In: Boston Studies in the Philosophy of Science, vol. 263, pp. 59–65. Springer, Dordrecht (2011)
Grantham, T.A.: Conceptualizing the (dis)unity of science. Philos. Sci. **71**(2), 133–155 (2004)
Kellert, S.: Borrowed Knowledge: Chaos Theory and the Challenge of Learning Across Disciplines. University of Chicago Press, Chicago (2008)
Lakatos, I.: History of science and its rational reconstructions. Boston Stud. Philos. Sci. **8**, 91–136 (1971)

Laudan, L.: Progress and Its Problems: Towards a Theory of Scientific Growth. University of California Press, Berkeley (1977)
Laudan, L.: Thoughts on HPS: 20 years later. Stud. Hist. Philos. Sci. **20**, 9–13 (1989)
Mäki, U.: Economics imperialism: concept and constraints. Philos. Soc. Sci. **39**, 351–380 (2009)
Maull, N.: Unifying science without reduction. Stud. Hist. Philos. Sci. **8**, 143–162 (1977)
Mitchell, S., Daston, L., Gigerenzer, G., Sesardic, N., Sloep, P.: The whys and hows of interdisciplinarity. In: Weingart, P., et al. (eds.) Human by Nature, pp. 103–150. Lawrence Erlbaum Associates, Mahwah (1997)
Nickles, T.: What is a problem that we may solve it? Synthese **47**, 85–118 (1981)
Nickles, T.: Heuristic appraisal: context of discovery or justification? In: Schikore, J., Steinle, F. (eds.) Revisiting Discovery and Justification, pp. 159–182. Springer, Berlin (2006)
Schickore, J.: More thoughts on HPS: another 20 years later. Perspect. Sci. **19**(4), 453–481 (2011)
Sherif, M., Sherif, C.W.: Interdisciplinary coordination as validity check: retrospect and prospects. In: Sherif, M., Sherif, C.W. (eds.) Interdisciplinary Relationships in the Social Sciences, pp. 3–20. Aldine Transaction, Piscataway (1969)
Thompson Klein, J.: Interdisciplinarity: History, Theory, and Practice. Wayne State Univeristy Press, Detroit (1990)
Thorén, H., Persson, J.: Philosophy of interdisciplinarity: problem-feeding, conceptual drift, and methodological migration. Phil. Sci. Arch. http://philsci-archive.pitt.edu/8670/ (2011)
Thorén, H., Persson, J.: The philosophy of interdisciplinarity: sustainability science and problem-feeding. J. Gen. Philos. Sci. **44**(2), 337–355 (2013)
Toulmin, S.: Human Understanding, vol. I. Claredon Press, Oxford (1972)
Wylie, A.: Discourse, practice, context: from HPS to interdisciplinary science studies. In: PSA: Proceedings of the Biennial Meeting of the Philosophy of Science Association 1994 (II), pp. 393–395 (1995)

Context-Dependent Anomalies and Strategies for Resolving Disagreement

A Case in Empirical Philosophy of Science

Douglas Allchin

Abstract The interpretation and analysis of anomalies is itself theory-dependent, as illustrated in the case of the ox phos debate in biochemistry in the 1960s. Here, the perceived threat of six anomalies to an existing research lineage depended on perspective, or Kuhnian paradigm. The ambiguous status of anomalies sharpens the problem of Kuhnian incommensurability. But analysis of the details of the historical case—one way to pursue an empirical philosophy of science—also indicate a possible solution. The asymmetric organization of multiple anomalies strongly indicated that disagreement had shifted from an intraparadigm to an interparadigm level, where modes of effective argument and use of evidence differ. This diagnostic awareness of the type of disagreement can orient discourse and allow investigators to develop and present evidence appropriately. I briefly extend the results of this historical case analysis to Darwin's synthesis and to gendered bias in craniology, to indicate the prospective generality of the analysis of anomaly asymmetry.

Keywords Empirical philosophy of science · Anomalies · Kuhn · Incommensurability · Error types · Strategies

---

An earlier version of this paper was presented as "Anomalies in Ox-Phos: Six of One Theory, a Half-Dozen of Another," at 2003 meeting of the International Society for the History, Philosophy and Social Studies of Biology in Guelph, Ontario. My appreciation to Lindley Darden and Kevin Elliot for fruitful discussion of that presentation.

D. Allchin (✉)
Minnesota Center for the Philosophy of Science, University of Minnesota, Minneapolis, MN 55455, USA
e-mail: allch001@umn.edu

D. Allchin
Science Studies Program, Århus University, Aarhus, Denmark

1 Introduction

How can history contribute to an empirical philosophy of science? In particular, how can one bridge the gap between abstractly normative and concretely descriptive accounts? Here, I offer a case with one prospective solution.

At one level, any methodological question about science is necessarily empirical: does the idealized method proposed by philosophers actually work in practice? In what contexts, or under what circumstances? The relation between history of science and philosophy of science has always been viewed as somewhat problematic, even if also fruitful (Brush 2007; Losee 1987; Nickles 1995). Nonetheless, several major efforts have effectively demonstrated the value of "testing" philosophical propositions through analysis of historical cases (Brush 2015; Donovan et al. 1988; Hull 1993 ; Losee 1972, 2005). Similarly, one might ask whether, based on history, the epistemological dimension of social norms envisioned by Merton (1973), Hull (1988), or Longino (1990) are, or can be, realized in practice (Jukola this volume). Indeed, good historical analysis may well shape an impression of what epistemological goals are achievable, or what one can realistically target. Empirical perspectives support a naturalized epistemology, sensitive to the abilities and limits of human cognition (Bechtel and Richardson 2010; Callebaut 1993; Wimsatt 2007). In these approaches, history provides the evidence for assessing the validity and scope of philosophical theories about how science should, can, or does function.

Another approach, which I explore here, is to adopt standard philosophical norms about scientific knowledge (consider such familiar benchmarks as reliability, simplicity, explanatory power, predictiveness, or novelty), while remaining uncommitted about the possible methods for achieving them in practice. Here, philosophy may offer epistemological, or normative, aims and justifications—the "whys". However, history answers the epistemic, or descriptive, questions—the "hows" of scientific practice (Losee 1972, 1987). That is, philosophy stipulates the ultimate values; nitty-gritty history, the proximal mechanisms. Product and process differ. For example, one may aim for consistency between theory and evidence. But in practice, experimental findings may not align with theoretical predictions. That is, anomalies may emerge. Such inconsistencies are ideally resolved. But philosophers generally do not prescribe how such anomalies are resolved. Through an analysis of history, however, and by documenting many examples of resolving anomalies in Mendelian genetics, Darden (1991) was able to generate a practical repertoire of potential strategies that might guide scientists on other occasions in the future. Similarly, Bechtel and Richardson (2010) acknowledged reductive explanation as a conceptual goal, but considered a large sample of historical cases in order to articulate just how scientists typically do this successfully in practice. The descriptive work of history makes the normative perspective of philosophy more complete and applicable.

To illustrate this approach further, I consider how the ox phos debate in cellular biochemistry in the 1960s might inform classic philosophical problems about Kuhnian paradigm shifts (for a fuller account, see Allchin 1991). This episode exemplifies well the type of dramatic theoretical and methodological gulfs or alternative gestalts described by Kuhn (1970; see also Hoyningen-Heune 1993; Allchin 1992, 1994; Weber 2002). In stormy rhetoric participants seemed (as informed by a retrospective view) to blindly talk past each other. Their discourse exhibited vividly the challenges of Kuhnian incommensurability where commitments to alternative problem fields differ and evidence could not be measured using comparable assumptions or benchmarks. Ultimately, the participants did resolve the disagreement after a decade and a half of debate, by redefining and differentiating the empirical domains, or scope, of the conflicting theories and their corresponding suites of experimental practices (Allchin 1994, 1996, 1997).

Kuhn maintained that interparadigm disagreement, aggravated by the challenge of incommensurability in discourse, is eventually resolved rationally, although he was not able to fully articulate just how, at least to the satisfaction of many skeptics and critics. How, indeed, does one interpret and resolve problematic interparadigm disagreements from the historically situated perspective of science-in-the-making? That is an empirical question, with important overtones at for general philosophical conceptions. The ultimate epistemic aim, here, may be achieved in part through concrete historical analysis. The proximal historical aim is to interpret how practitioners could transition from apparently irreconcilably conflicting views to acceptably complementary views. Namely, once the debate had begun, how could researchers interact productively to resolve it?

In the case of the ox phos controversy, viewed retrospectively, one particular problem was especially noteworthy. Earlier, I characterized how effective evidence-based argumentation differs for intraparadigm versus interparadigm disagreement (Allchin 1991, 1992, 1994). For example, crucial either-or tests may be possible within a paradigm, where assumptions and background knowledge are stable. But where problem fields and assumptions diverge, as in an interparadigm context, one must rely more on demonstrations, which merely display the explanatory power of a theory without decisively ruling our specific alternatives (Allchin 1994; Robinson 1984). Throughout much of the ox-phos debate, however, chemists engaged in intraparadigmatic arguments, trying (unsuccessfully) to resolve interparadigmamic discord. By misframing the discourse, and relying on implicit assumptions that were not shared, they tended to talk past each other. While one may easily see this in retrospect, it is less clear how participants in the midst of such historical developments may recognize the circumstances. This practical problem, while based on a philosophical understanding, calls for empirical analysis of history. How does someone know when disagreement has shifted from an intraparadigm to an interparadigm level, changing the terms of evidential argumentation? What diagnostic clues are available?

As one examines the case closely with these factors in mind, one finds that the dire sketch Kuhn provided of conceptual change was, ironically, somewhat optimistic. He regarded anomalies as well defined, able to leverage a "crisis." In the ox phos case, however, the interpretation or analysis of anomalies itself depended on theoretical context (Gilbert and Mulkay 1984). That is, an anomaly for one scientist was not necessarily the same anomaly for another—and may not have seemed anomalous at all. Philosophically, one might find here occasion to further characterize the difficulties of incommensurability, or to criticize and revise Kuhn's model. That would treat history as evidence for informing philosophical theories (first approach above). However, this is not my primary goal. Rather, history can also afford a more active role in informing and enriching philosophical perspectives. One can analyze the history and discover—not test—how the disagreement was, ultimately, resolved. In addition, this understanding could help inform science in practice. One can solve the Kuhnian problem of incommensurability and inter-paradigm disagreement empirically, not conceptually. Still, an answer, once discerned, can certainly be framed (retrospectively) with a philosophical flourish, deepening our abstract conceptual understanding of Kuhnian-type episodes in science.

2 Interpreting the Anomalies of Ox Phos

Let us enter the case in 1961.[1] Hans Krebs has elucidated the reactions of the citric acid cycle. Fritz Lipmann has described the central role of phosphate bonds, notably in ATP, as an energy carrier in the cell. David Keilin has helped identify the cytochrome chain that transforms energy from the Krebs cycle to ATP. For the last decade, research has focused on deciphering these final energy reactions that use oxygen and produce ATP: oxidative phosphorylation, or ox phos. The general consensus is that there are more, yet unknown, chemical reactions with many high-energy intermediate compounds along the reaction pathway. Yet in the eight years since they were formally proposed in 1953, no one has found them.

At this time, Peter Mitchell introduced a remarkably different theory, which would ultimately earn him a Nobel Prize in 1978: what he called the chemiosmotic hypothesis. In his original 1961 paper, in a deceptively modest four column-inches of text citing twenty articles, Mitchell presented six anomalies: "facts," he said, "... that are generally acknowledged to be difficult to reconcile with this orthodox (chemical) view" (1961, 144). It was almost a textbook definition of anomalies. These six anomalies, Mitchell suggested, collectively prompted doubt in the reigning concepts about the high-energy intermediates, and instead supported his alternative interpretation, based on electrochemical membrane gradients. What interests us most, however, is not how other chemists weighed the evidence Mitchell presented or considered the relative merit of alternative theories. Rather, of interest is how they first interpreted, or gave meaning to, these experimental "facts"—and how this makes philosophical thinking about anomalies more complex.

---

[1] For a more complete account of the entire ox phos episode, see Allchin (1991, 1997), and Weber (1991).

First, Mitchell noted that loss of the ATP product on one side of the mitochondrial membrane led to changes in the equilibrium of the reactions on the other side of the membrane. Mitchell contended that moving hydrogen ions across the membrane was central to the energy reactions—and here he emphasized how his conception could explain this particular effect, bridging the two sides of the membrane. But while chemists acknowledged this fact, they did not see it as threatening their view. They saw ox-phos, like all chemical reactions, as reversible. When one uses the product, equilibrium shifts. There was no broken expectation, no inadequate explanation. No anomaly, here, at least.

Second, Mitchell noted, the proposed high-energy intermediates of the reaction series were "elusive to identification". In classic scientific understatement, he had implied, of course, that there were no intermediates at all. Rather, the intermediate energy stage was a build-up of protons outside the membrane: an electrochemical pH gradient. Those studying ox phos were arguing about whether such intermediates were phosphorylated, or whether there was a second non-phosphorylated intermediate, so Mitchell's claim seemed to betray a fundamental confusion. Moreover, from recent reports, biochemists seemed on the verge of isolating the intermediates. They were likely short-lived and thus hard to isolate experimentally, especially if embedded in the membrane. This was a technical puzzle so typical of Kuhnian normal science, not a theoretical failure—and certainly not epistemically threatening (Allchin 1997).

Third, Mitchell noted, structurally intact membranes seemed essential. For Mitchell, the membrane preserved the pH energy gradient. Here, chemists did consider this problematic—but only experimentally. The conventional research, epitomized in the work of Krebs, Lipmann and others, targeted enzymes in aqueous solutions. The ox-phos components, however, were located in the mitochondrial membrane, a hydrophobic (or oil-like) environment. Researchers could not isolate the components while still functional. For biochemists, the challenge was largely another technical puzzle of normal science: to discover how to isolate enzymes intact from membrane-like structures. Later, Lehninger (1960) viewed the membrane more positively: "There may be a biological necessity for structural organization of these catalysts in a moderately rigid, geometrically organized constellation in the membrane." The membrane might hold enzymes in close proximity and proper orientation. The implied remedy, as before, was to search experimentally for ways to prepare such complex membrane-bound structures. The same acknowledged "fact"—the structural integrity of membranes—had two quite different meanings: one as a technical puzzle, the other as threatening theory and the way of doing ox phos science.

Fourth, Mitchell noted that many compounds interfered with ox phos, but they seemed to share no specific chemical characteristic. Mitchell noted, however, that these compounds were all soluble in the membrane's oil-like environment. They could thus enter the membrane and transport protons (or other charged particles), dissipating the pH gradient. For chemists, the solubility could certainly explain how the compounds entered the membrane. But understanding how they worked required more specific elucidation of their structure. Mitchell seemed to miss the critical features, which might not be known until all the reactions and their enzymes had been studied. Nor did anything dictate one common mechanism for all the chemicals.

Fifth on Mitchell's list: mitochondria would swell and shrink during ox-phos. According to the chemiosmotic view, the movement of various ions caused the corresponding osmotic movement of water. While such osmotic effects were not uncommon, they were more familiar to the lipid biochemists who studied membranes. Biochemists studying energy-related reactions focused primarily on enzymes and protein chemistry. Osmotic phenomena fell outside their concerns. Swelling might occur incidentally, as a by-product, but hardly seemed relevant to how the enzymes functioned. Here, Mitchell and the other chemists addressed different potentially relevant variables.

Last among Mitchell's list of anomalies: reactants and products did not always exhibit integer ratios. When studying chemistry, we all learned to balance chemical equations. Reactants relate to products in whole number ratios. Chemists observed that this "rule" was occasionally broken for mitochondria. For Mitchell, even if the reactions creating the gradient followed exact ratios, the pH gradient of the membrane could "leak" any amount. Other chemists acknowledged, for their part, that intermediate products might be used in other reactions, altering observed ratios. The uneven ratios, so commonly observed, reflected experimental static, or noise, not meaningful signal. Technical mastery would eventually dissolve this artifact—another puzzle for normal science. Once again, the chemists isolated Mitchell's "anomaly" to experimental methods, not theoretical concepts (Allchin 1997).

So, there were six anomalies. All could agree in 1961 about the basic "facts" or experimental observations. Yet where Mitchell saw many fundamental counter instances and explanatory flaws, chemists perceived only a handful of familiar technical puzzles and sometimes no problem at all. Mitchell saw the anomalies as evidence for a revolutionary new theory. Other chemists saw only Kuhnian normal science. Mitchell's anomalies were only anomalous using the chemiosmotic perspective as an interpretive guide. The meaning of the six anomalies was context-dependent. That is, while all agreed there was a latent error inherent in the accepted experimental results, they disagreed about how to localize, and thus clearly identify, that error. Of course, this should surprise no one. Anomalies, like any observation, may be theory-laden, or interpreted contextually. Accordingly, Lightman and Gingerich (1992), observed that anomalies do not begin with internal contradictions, but rather when a new paradigm introduces an alternative perspective that exposes them. The meaning, not merely the acknowledgement, of anomalies seems theory-dependent.

The history thus indicates how Kuhn's initial philosophical conception (although itself based on historical study) was rough or incomplete. Empirical historical analysis refines the philosophical concept. Here, the problem of incommensurability becomes even worse. According to Kuhn, an accumulation of anomalies leads to crisis. They reveal weaknesses in the paradigm that eventually lead to questioning it and developing a successor. In the ox-phos case, however, the view from within the established paradigm seemed to eclipse the type of awareness that Kuhn suggested becomes inevitable. Worse, perhaps, where Mitchell saw the opening of a new paradigm, other chemists saw only the continuity of normal science and puzzle-solving. Indeed, the divergent views seem to epitomize Kuhnian incommen-surability, but an incommensurability based on problem fields and views of relevance more than on linguistic references or communication woes (Allchin 1990). How, then, can anomalies lead to scientific change? How can one correct an error if researchers are blinded to its "meaning"—that is, if the interpretation of anomalies is itself theory-laden? Mere philosophical reflection does not necessarily solve the problem. History—the empirical dimension—has an additional role in profiling the solution, to which my discussion now turns.

3 Resolving Disagreement About Anomalies

Darden (1991) has suggested a set of strategies for resolving anomalies. They are not normative "methods," or algorithmic rules, in the conventional sense. They are possible solutions to explore. They are "strategies" derived from a more or less descriptive historical analysis, then formalized in a philosophical perspective. They do not guarantee results, but provide guidance whose potential value is warranted by historical experience. Darden's strategies on anomalies, however, were oriented exclusively to theory change, for cases where the "problem" is identifiably theoretical. In the ox phos case, as just noted, some chemists saw the problem as conceptual, others as experimental. One needs a broader perspective here.

As exemplified in the ox-phos case, one cannot always immediately isolate an individual anomaly unambiguously. Yet if one assumes that every anomaly exposes a latent "error" to be remedied, then to isolate anomalies or resolve disagreement, one may profit from a complete inventory of generalized error types. In contrast to Darden's focus on revising theories only, one may find that error types range from the material or experimental to the conceptual or discursive-social (Allchin 2001). The appearance of an anomaly does not itself indicate whether to localize the problem in the lab, in the theory, in cognitive or cultural biases, or in some other element of scientific practice. The problem in the ox phos case was how researchers, despite their divergent interpretations of the relevant error type(s), could communicate and argue effectively about them. How could Mitchell (or others) frame their evidence to be persuasive?

Here, the detailed historical perspective highlights an important clue in the pattern of the anomalies themselves. This was distinct from how each was interpreted. That is, the anomalies have a character as an ensemble, rather than individually. From Mitchell's chemiosmotic perspective, they formed a unified syndrome. They all implicated the relevance of the membrane. The flaws, as Mitchell framed them, were systematic. They functioned together as a half-dozen anomalies. From the extant perspective in ox-phos, on the other hand, these were six separate anomalies. In this case, six of one was not the same as a half-dozen of the other. This distinctive asymmetry was critical. While it did not provide a definitive solution, it showed the path to a solution. It indicated how to address the underlying disagreement.

What did the asymmetry mean? Today, in retrospect, we might say one could weigh the two theories by applying a standard philosophical norm of simplicity, consilience, coherence, or conceptual economy, and decide that adopting the new theory solved everything all at once (Janssen 2001). Namely, using the age-old Occam's razor, the chemiosmotic perspective was the clear "winner". Indeed, many researchers were impressed by the coherence of the chemiosmotic gestalt and began to entertain it seriously or reorient their research trajectories (Robinson 1984). Here, then, using the historical analysis, one could supplement Darden's catalog: namely, use a meta-analysis of multiple anomalies or error-types to identify a common error. This strategy echoes one sketched by Glymour (1980) for a more conventional hypothetico-deductive (logical) framework. Namely, when multiple observations or results do not match separate theoretical predictions, check shared boundary conditions or auxiliary hypotheses supporting those predictions as probably incorrect. Just as independent observations or lines of reasoning from multiple sources may provide robust support for a particular conclusion, so too they may indicate a robust weakness, vulnerability, or error (Wimsatt 2007, pp. 43-74). Thus, a potential strategy, exhibited through an empirical analysis of this case, might be: "Search for an intersection of prospective error types among many anomalies." In this view, a half-dozen anomalies would be inherently more informative than six.—And perhaps decisive.

However, the fully empirical approach I am profiling proceeds differently. One must work philosophically from within the historical perspective, or science-in-the-making (Latour 1987). Namely, philosophical analysis can be biased by retrospect. One cannot fruitfully trump the situated perspectives of the researchers. In 1961, the evidence is not yet fully in. Mitchell could be wrong. Searching for a common root error is merely a strategy, not a final evaluative judgment, or normative rule. Our analysis must thus focus instead on the discursive dimension. How were the different perspectives reconciled through further evidence? In this case, researchers needed to know how to present their findings effectively for others to understand, and for them to have persuasive merit.

While Mitchell did not necessarily resolve all the anomalies at the outset, he did, nonetheless, dramatically change the discursive landscape. He had shown how the anomalies could be related. The chemiosmotic view resolved all the anomalies at once, by adopting a new theory, or conceptual gestalt (as described above). The conventional chemist who resolved one anomaly, still had five others to resolve. For example, showing that the membrane functioned as scaffolding for protein interaction would not thereby solve the anomaly of the missing intermediates, and vice versa. Piecemeal solutions for each anomaly no longer sufficed.

The six/half-dozen asymmetry was the critical contextual signal. Its significance was in indicating that discourse had shifted from intraparadigm to interparadigm comparisons. It did not yet resolve the disagreement. When Mitchell showed a plausible role for the membrane in all cases, he essentially destabilized the background assumptions that had guided earlier experimental reasoning and interpretation. Those assumptions could no longer be regarded as unproblematically justified. The asymmetric stacking of anomalies reflected this altered epistemic environment.

Returning to the historical perspective, how did this shift affect the researcher? Generally, when two theoretical alternatives present themselves, an investigator hopes to test them against each other under controlled conditions, isolating the variable in question against a stable background. However, in the context of contrasting paradigms, one can no longer make such narrow parallel comparisons. Mitchell could present experimental evidence that supported his view, but not that simultaneously excluded the chemists' interpretations as "wrong". Indeed, his criticisms fell relatively flat because chemists felt no need in 1961 to abandon their own interpretations. How could he present evidence, then, for the integrated nature of the six anomalies? Mitchell and others had to show, or demonstrate, that the chemiosmotic perspective was cogent and fruitful, and solved relevant problems (for more details, see Allchin 1992). The focus becomes experimental demonstrations, without undo concern for explicit comparisons or discounting of alternatives. In an interparadigm context, the appropriate strategy is demonstration. The asymmetric stacking of anomalies—six of one, a half-dozen of another—was essentially a diagnostic signal of, rather than a particular solution to, the shift in argumentation and experimental strategy to the interparadigm level. The framing of this significant diagnostic signal is the concrete outcome of an empirical historical approach in this case. This strategy was not obvious from a purely abstract perspective, I contend. Nonetheless, it emerges from a detailed analysis sensitive to historical perspective.

4 Conclusion

The case of asymmetry in anomalies in ox-phos, between six-of-one and a-half-dozen-of-the-other, then, may serve to illustrate a particular fruitful use of history in an empirical philosophy of science. For example, historical analysis may provide important specific "hows" where the philosophical "whys" are already established. The analysis may yield scientifically fruitful strategies, sensitive to context, such as, "When the interpretation of multiple anomalies differ (some viewing them as independent and others as conceptually unified), assume interparadigm discourse and adopt a strategy of experimental demonstration." Here, history has a creative role in developing—not merely assessing or contextualizing—philosophical principles. That is, the ox phos case helps illustrate how historical analysis can refine, and possibly revise, philosophical concepts; how history can go beyond conventional philosophical norms by articulating them in authentic scientific practice; and, most importantly, how history can help profile research strategies. As demonstrated in this case, history can contribute to the middle zone between abstract philosophical norms and concrete historical descriptions, where the "hows" are as important to scientists as the "whys."

References

Allchin, D.: Paradigms, populations and problem fields: approaches to disagreement. PSA **1990** (1), 53–66 (1990)

Allchin, D.: Resolving disagreement in science. Ph.D. dissertation, University of Chicago, Chicago (1991)

Allchin, D.: How do you falsify a question? Crucial tests versus crucial demonstrations. PSA **1992** (1), 74–88 (1992)

Allchin, D.: The super-bowl and the ox-phos controversy: winner-take-all competition in philosophy of science. PSA **1994**(1), 22–33 (1994)

Allchin, D.: Cellular and theoretical chimeras: piecing together how cells process energy. Stud. Hist. Philos. Sci. **27**, 31–41 (1996)

Allchin, D.: A 20th-century phlogiston: constructing error and differentiating domains. Perspect. Sci. **5**, 81–127 (1997)

Allchin, D.: Error types. Perspect. Sci. **9**, 38–59 (2001)

Bechtel, W., Richardson, R.: Discovering Complexity. IT Press, Cambridge (2010)

Brush, S.G.: Suggestions for the study of science. In: Gavroglu, K., Renn, J. (eds.) Positioning the History of Science [essays in honor of S. S. Schweber]. Boston Studies in the Philosophy of Science, vol. 248, pp. 13–25 (2007)

Brush, S.G.: Making 20th Century Science: How Theories Became Knowledge. Oxford Unviersity Press, New York (2015)

Callebaut, W.: Taking the Naturalistic Turn: How the Real Philosophy of Science is Done. University of Chicago Press, Chicago (1993)

Darden, L.: Theory Change in Science: Strategies form Mendelian Genetics. Oxford University Press, Oxford (1991)

Donovan, A., Laudan, L., Laudan, R.: Scrutinizing Science: Empirical Studies of Scientific Change. Springer, Berlin (1988)

Gilbert, G.N., Mulkay, M.: Opening Pandora's Box: A Sociological Analysis of Scientists' Discourse. Cambridge University Press, Cambridge (1984)

Glymour, C.: Theory and Evidence. Princeton University Press, Princeton (1980)

Hoyningen-Heune, P.: Restructuring Scientific Revolutions. University of Chicago Press, Chicago (1993)

Hull, D.: Science as a Process. University of Chicago Press, Chicago (1988)

Hull, D. Testing philosophical claims about science. In PSA 1992, vol. 2, D. Hull, M. Forbes and K. Okruhlik (eds.), East Lansing, MI: Philosophy of Science Association, pp. 468-475 (1993)

Janssen, M.: COI stories. Perspect. Sci. **10**, 457–522 (2001)

Latour, B.: Science in Action. Harvard University Press, Cambridge (1987)

Lehninger, A.L.: Oxidative phosphorylation in submitochondrial systems. Fed. Proc. **19**, 952–962 (1960)

Lightman, A., Gingerich, O.: When do anomalies begin? Science **55**, 690–695 (1992)

Longino, H.: Science as Social Knowledge: Values and Objectivity in Scientific Inquiry. Princeton University Press, Princeton, NJ (1990)

Losee, J.: A Historical Introduction to the Philosophy of Science. Oxford University Press, Oxford (1972)

Losee, J.: Philosophy of Science and Historical Enquiry. Oxford University Press, Oxford (1987)

Losee, J.: Theories on the Scrap Heap: Scientists and Philosophers on the Falsification, Rejection and Replacement of Theories. Pittsburgh University Press, Pittsburgh (2005)

Merton, R.K.: The normative structure of science. In: The Sociology of Science, pp. 267–78. University of Chicago Press, Chicago (1973)

Mitchell, P.: Coupling of phosphorylation to electron and hydrogen transfer by a chemi-osmotic type of mechanism. Nature **191**, 144–148 (1961)

Nickles, T.: Philosophy of science and history of science. Osiris **10**, 139–163 (1995)

Robinson, J.D.: The chemiosmotic hypothesis of energy coupling and the path of scientific opportunity. Perspect. Biol. Med. **27**, 367–383 (1984)

Weber, B.: Glynn and the conceptual development of the chemiosmotic theory: a retrospective and prospective view. Biosci. Rep. **11**, 577–617 (1991)

Weber, M.: Incommensurability and theory comparison in experimental biology. Biol. Philos. **17**, 155–169 (2002)

Wimsatt, W.C.: Re-Engineering Philosophy for Limited Beings. Harvard University Press, Cambridge (2007)

MIX
Papier aus verantwortungsvollen Quellen
Paper from responsible sources
FSC® C105338

If you have any concerns about our products,
you can contact us on
ProductSafety@springernature.com

In case Publisher is established outside the EU,
the EU authorized representative is:
**Springer Nature Customer Service Center GmbH
Europaplatz 3, 69115 Heidelberg, Germany**

Printed by Libri Plureos GmbH
in Hamburg, Germany